EveryDay Geography
of the United States

EveryDay Geography of the United States

A Lively Look at the Land, Climate, People & History of the 50 States

Barbara Fifer

Illustrations by Tim Steger

BLACK DOG
& LEVENTHAL
PUBLISHERS
NEW YORK

Published by
Black Dog & Leventhal Publishers
151 West 19th Street
New York, NY 10011

Distributed by
Workman Publishing Company
708 Broadway
New York, NY 10003

Text by Barbara Fifer
Illustrations by Tim Steger

Interior design by Cindy Joy
Jacket design by Sheila Hart

Manufactured in the United States of America.

ISBN: 1-57912-325-2

g f e d c b a

CONTENTS

Introduction 7
Alabama 15
Alaska 18
Arizona 21
Arkansas 24
California 27
Colorado 30
Connecticut 33
Delaware 36
Florida 39
Georgia. 42
Hawaii 45
Idaho 48
Illinois 51
Indiana 54
Iowa 57
Kansas 60
Kentucky 63
Louisiana 66
Maine 69
Maryland 72
Massachusetts 75
Michigan 78
Minnesota 81
Mississippi 84
Missouri 87
Montana 90

Nebraska. 93
Nevada 96
New Hampshire 99
New Jersey 102
New Mexico 105
New York 108
North Carolina 111
North Dakota 114
Ohio 117
Oklahoma 120
Oregon 123
Pennsylvania 126
Rhode Island 129
South Carolina 132
South Dakota 135
Tennessee 138
Texas 141
Utah 144
Vermont 147
Virginia 150
Washington 153
West Virginia 156
Wisconsin 159
Wyoming 162
District of Columbia 165
Territories 167
Index 171

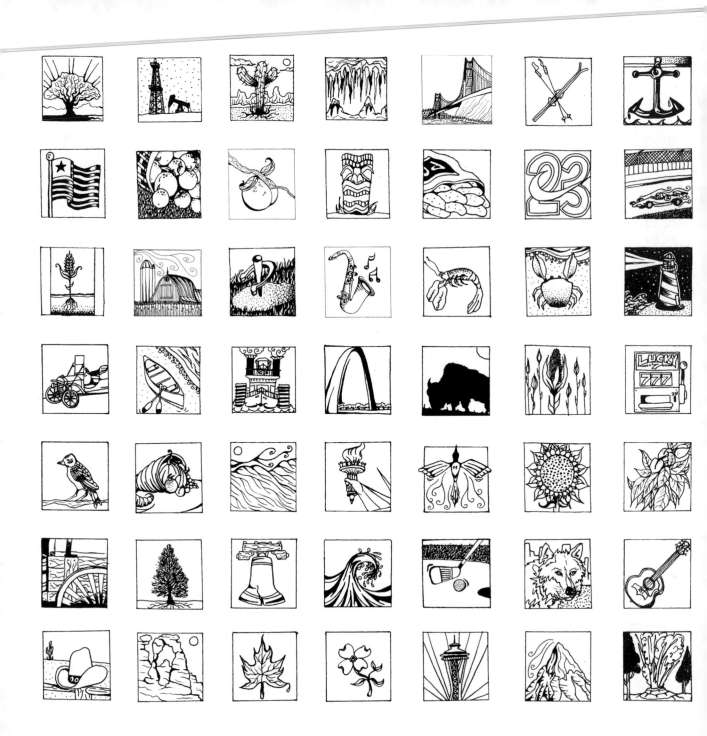

THE UNITED STATES OF AMERICA

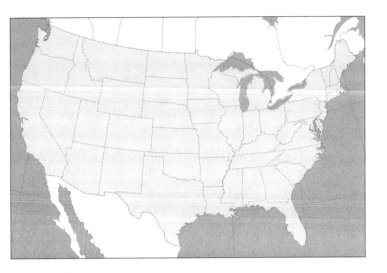

square miles: 3,537,438
population: 281,421,906
density: 79.6 people per square mile
capital: Washington, D.C.
largest city: Los Angeles/Long Beach, California
independence: July 4, 1776
motto: In God We Trust
bird: bald eagle
flower: rose

Land

The United States of America is the world's fourth-largest nation physically, after Russia, Canada, and China. On the lower half of North America it spreads from 3,700 miles of Atlantic Ocean/Gulf of Mexico coastline on the east to 1,300 miles of Pacific Ocean coast on the west. It also includes Alaska to the northwest and Hawaii more than 2,000 miles west. Since Alaska joined the Union, the forty-eight contiguous (touching) states often are called the lower 48. The U.S. shares unfortified borders with Canada north of the lower 48 and east of Alaska (5,500 miles), and with Mexico to the south (1,933 miles).

7

Part of the northern border extends through the five Great Lakes and another part follows the St. Lawrence River to the east. A portion of the southern border is the Rio Grande. The Ohio/Missouri/Mississippi river system is the nation's largest. Along the Gulf of Mexico and Atlantic Ocean to New York, the low-lying coastal plain reaches as far inland as 200 miles. Its shores are mostly sandy or marshy. North of New York, New England is mountainous, with a mostly rocky coastline.

Inland from the coastal plain is the Fall Line, where rivers flowing to the Atlantic drop in waterfalls. This point is as far upriver as ships can travel. Many major cities were built on the Fall Line, including Philadelphia, and Washington, D.C.

West of the Fall Line, the Piedmont Plateau has elevations of 300' to 1,000', then gives way to the abrupt rise of the Appalachian Mountains, which roughly parallel the Atlantic coast from Alabama into Canada.

Extending as far as 1,000 miles west of the Appalachians, central lowland and the Great Plains reach from Canada to the gulf coastal plain. Some mountains rise from this area: the Mesabi Range, the Ozarks, the Ouachita, and the Black Hills. The plains' elevation gradually rises toward the west and the Rockies.

The Rocky Mountains are the lower 48's highest point, extending from Canada to New Mexico. Their crest is the Continental Divide: Rainwater falling on the east side feeds rivers flowing to the Gulf of Mexico, and water on the west flows toward the Pacific Ocean. High-elevation plateaus lie west of the Rockies, broken by deep canyons cut by the Snake (Hells Canyon), Columbia (Columbia Gorge), and Colorado (Grand Canyon) rivers.

West of the northern plateau is the Great Basin, semiarid land where Utah's Great Salt Lake is the remnant of an ancient inland sea. South of there, in Nevada and California, low land holds hot, arid Death Valley (the nation's lowest point at -282').

The volcanic Cascade Mountains extend from Canada to northern California, and south of them are California's Sierra Nevada. To their west, a wide basin holds the Great Valley in California and the Willamette River valley and the Puget Trough in Oregon and Washington.

From Alaska to northern California, the Coast Ranges provide rocky ocean cliffs with some sandy coves. South of them, beaches are sandy. The Pacific does not have an inland coastal plain like the Atlantic.

Landforms of Alaska and Hawaii are described in their individual entries.

People & History

The United States is the world's third most populous nation, after China and India. It is a nation of immigrants on a continent of immigrants.

No skeletons predating those of Homo sapiens (modern humans) have been found in North or South America, as they have been on other continents. One theory of how people arrived on these continents is that they came from Asia about 11,500 years ago over a land bridge to Alaska. At the end of the last Ice Age, glaciers held much of the oceans' water, and sea levels were lower. A newer theory suggests that people came from both Asia and Europe by boats, traveling along the edges of glaciers rather than over the open sea. This theory is an attempt to explain 15,000-year-old artifacts and skeletons without Asian characteristics.

Thousands of different cultures had evolved in the Americas by the time European explorers reached the Caribbean Sea in 1492. By then, an estimated 40 million people lived in North America and the Caribbean. Because the first explorer—Christopher Columbus—thought he had reached islands off India, he called the people Indians. Today they are called either Native Americans, American Indians, or Indigenous Peoples.

Europeans brought diseases that had not existed in the New World, such as smallpox. Over the centuries of initial contact between Europeans and Native Americans, such illnesses killed many more Indigenous Peoples than warfare with the settlers.

European nations fought to control North America from the 1500s to the 1700s. In general, England created colonies on the east coast; Spain on the west coast, in the southeast, and southwest; and France moved along and out from the Mississippi River.

England's thirteen colonies became the core of the future United States. They are Connecticut, Delaware, Georgia, Maryland, Massachusetts, New Hampshire, New Jersey, New York, North Carolina, Pennsylvania, Rhode Island, South Carolina, and Virginia. By the 18th century, their residents were from many European countries and included African slaves.

By the mid-18th century, the colonies wanted a government that better fit their lives. They also wanted to invest in their own country rather than pay taxes to a far-off nation. In public and also secret meetings, people talked about a new kind of government, not controlled by a king. They wanted democracy, where citizens voted to make their own laws. In its earliest form, this democracy defined citizens as white men with some property.

Britain tried to stop this movement, and began the Revolutionary War by attacking Massachusetts militiamen in 1775. On July 4, 1776, the thirteen colonies adopted a formal Declaration of Independence. The Revolutionary War continued until American victory in 1781, and a treaty between the two nations was signed in 1783.

Americans at first tried a confederation of states, but in 1787 leaders wrote the Constitution that gave some powers to a central–or federal–government, and some powers to the states. The thirteen states took three years to ratify, or accept, the Constitution. Rhode Island refused to sign until the first ten amendments were ready to be added. Called the Bill of Rights, they guarantee citizens certain protections and privileges. When Rhode Island ratified the Constitution on May 29, 1790, the United States of America truly existed.

From the nation's beginning, slavery caused great disagreement. Northern states had banned it, but Southern states depended upon slaves in their plantation economy.

Americans had begun settling west of the colonies, but after the Revolution more followed. Indians both fought and moved farther west themselves. By 1803, four more states had entered the Union.

That year, President Thomas Jefferson nearly doubled the nation's size when he bought the Louisiana Territory from France. The Louisiana Purchase covered all land drained by the Missouri River system. This meant that the U.S. now controlled land east of the Continental Divide.

Northern states were prospering with factories, international shipping, and trade. Southern states depended on agricultural products, many sold abroad. But France's Napoleon wanted to conquer Britain. To weaken it, French ships blockaded its ocean trade which included U.S. ships. Britain also stopped American ships, often forcing U.S. crews to sail for the British navy. Its economy suffering, the U. S. declared war on Britain in 1812. When the war ended in 1814, both sides claimed victory, but the U.S. had turned away British invaders.

By 1819, twenty-two states were in the Union, half of them slave and half free. Then Missouri, which allowed slavery and was west of the Mason-Dixon Line (traditional division between slave and free states), applied for statehood. Congress created the Missouri Compromise in 1820. It admitted Maine, a free state, that year, and authorized Missouri to prepare for statehood. The compromise also banned slavery from future states beyond Missouri's northern border. The 1854 Kansas-Nebraska Act changed the ban, allowing future western states to vote on slavery.

Americans continued to move west and take Indian lands. By 1830, an area in the southern plains was set aside as Indian Territory and most native people east of the Mississippi were forced to move there. In twenty years, this area was chipped away by white settlers, leaving Indians' land reduced to about the borders of present Oklahoma.

A several-year depression, the Panic of 1837, temporarily slowed America's growth. Congress had allowed anyone to buy public land whether they lived on it or not. People were borrowing money to buy land they hoped to resell at a higher price, and a large number did not repay their loans, causing many banks to fail.

The Republic of Texas joined the Union as the 28th state in 1845. Mexico, which did not agree to the Rio Grande as the border, had warned the U.S. not to allow Texas to become a state. The U.S. claimed Mexico owed it millions of dollars for American property seized and lives taken since Mexican independence in 1821. Also, Americans had begun moving into Mexican territory, which later became New Mexico, Arizona, and California. A popular view called Manifest Destiny said the U.S. must expand to the Pacific Ocean.

The Mexican War began with a Mexican attack on Texas in 1846. With a U.S. victory in 1848 came ownership of most of today's Southwest.

Just ten days before the 1848 treaty ending that war was signed, a carpenter, while building a sawmill, found gold in California. During 1849, tens of thousands of prospectors—nicknamed forty-niners—and businesspeople poured into California. That year, its population grew from 15,000 to 100,000-plus.

In the 1840s, agricultural settlers had begun moving by wagon train to Oregon. The 1850s saw wagon trains to Washington territory. Now white settlement moved east from the West Coast as well as west from the Midwest.

States admitted to the Union from 1836 through 1848 had continued the half-slave/half-free balance, which ended when California became a state in 1850. Congress passed the Compromise of 1850, which provided stricter laws for capturing runaway slaves, banning slavery in the District of Columbia, making California a free state but leaving the decision of slavery to voters in the Utah and New Mexico Territories. These laws were a temporary solution, for tensions about slavery kept mounting.

The Underground Railroad helped slaves escape to free states and Canada. It was a loose network of people who guided fleeing slaves to the North, overland by night and hiding them in safe houses by day. Anti-slavery groups in the North became more popular. Meanwhile, Southerners argued that slavery, among other issues, should be determined by individual states. The economics of the industrializing North and the agricultural South was one of the main divisions in the issue of slavery.

In February 1861, six southeastern states seceded from the Union, forming the Confederate States of America. Five more states joined, but the Confederacy claimed thirteen members because it set up governments inside Kentucky and Missouri, which did not secede. When Confederate troops attacked Fort

Sumter, South Carolina, in 1861, the Civil War began. It lasted four years, was fought mostly on Southern soil (destroying farms, railroads, and towns), and was this nation's bloodiest war. Battles frequently took the lives of one-fourth of the participants, but twice as many men died from illness as from battle.

On New Year's Day 1863, President Abraham Lincoln declared in the Emancipaton Proclamation that all slaves in seceded states were free. In 1865, Congress passed the 13th Amendment to the Constitution, banning slavery throughout the U.S.

After the North won the war in 1865, a period called Reconstruction followed, which lasted until 1877. It rebuilt the South, created loyal state governments, and readmitted former Confederate states to the union. The issue of what rights belonged to the states was again debated nationally.

Gold strikes in the Rocky Mountains drew prospectors to Colorado just before the Civil War, and to the northern Rockies during the conflict. Building of the first transcontinental railroad also started—east from California in 1863 and west from Nebraska in 1865. The era's fastest transportation, railroads, would tie the West's natural resources to the East's markets. In 1869, the lines met in Utah.

Farmers and ranchers moved onto the Great Plains in large numbers when the Homestead Act was made law in 1862, which enabled farmers to own their land after five years of farming a plot. By 1900, half a million people had benefited, and population west of the Mississippi had nearly tripled.

Attention turned to subduing Indian nations of the north and south plains. Since obtaining horses in the 1600s, Plains Indians had been nomadic, moving their villages to hunt buffalo. Ranches with fences and cattle did not blend in with this lifestyle. By 1890, most of the buffalo had been killed, and the Indians moved onto reservations. That same year, the Census Bureau announced that the nation no longer had a frontier.

A depression from 1893 to 1897 slowed growth, but the U.S. swiftly recovered. Open immigration until 1924 helped increase U.S. population.

When World War I began in Europe in 1914, many Americans were against joining the defense of the Allies, France, and Britain. The U.S. finally entered the war in 1917 and celebrated the Allied victory in 1919 with the Treaty of Versailles.

The 1920s brought women's suffrage, automobiles, radio, and motion pictures. As of the 1920 census, and for the first time in history, most U.S. residents lived in cities. Food prices were low, and farmers suffered. Federal prohibition of making and selling alcoholic beverages, which lasted from 1920 to 1933, enriched gangsters, who took over the industry. Buying stock became a national craze, and many people bought with borrowed money. A stock market crash in 1929 began the decade-long Great Depression.

At its worst, in 1933, one-fourth of American workers were jobless. Federal programs were enacted to employ some people and support agricultural prices. Many public works, such as dams, were constructed. Dry climate conditions collided with past overfarming in the mid-1930s, and tons of midwestern topsoil blew away. People called the agricultural heartland the Dust Bowl.

World War II began in Europe in 1939. Factory and farm production put America back to work. Late in 1941, Japan attacked Pearl Harbor in Hawaii and the U.S. declared war. Germany and Italy, allies of Japan, declared war on the U.S. a few days later, so the U.S. was at war in Europe as well as in the Pacific. During the nation's largest war effort, factories produced war materials, and civilians endured rationing of food, gasoline, and other items.

The war in Europe ended with Germany's surrender in May 1945; the Pacific war ended with Japan's surrender in August after the U.S. had dropped two newly invented atomic bombs that destroyed major cities.

Despite U.S. involvement in the Korean War (1950–1953), the 1950s were a prosperous decade as veterans of World War II started families and built homes; suburbs began to grow around cities. The U.S. and its allies worried about the spread of communism as the Soviet Union took on satellite nations in eastern Europe. The Cold War referred to the resulting international tension and arms buildup.

Preventing communism's spread caused the U.S. to support anti-communists in Southeast Asia, especially South Vietnam, beginning in the early 1960s. Despite increasing numbers of U.S. and allied nations' soldiers and years of peace negotiations, the war between North and South Vietnam stalemated. It gradually became more unpopular with the American public. In 1972, the U.S. withdrew, and in 1975 Saigon fell to the North Vietnamese.

In the 1950s and 1960s, the civil rights movement worked for equal rights for African Americans, denied in the South since Reconstruction. Participants used boycotts, demonstrations, and lawsuits, and sometimes were met by violence. Their struggle finally resulted in the passage of the federal Civil Rights Act of 1964.

An economic recession in the early 1980s helped cause many factory closings. Older factories, especially in the northeast, were abandoned rather than updated. "High-tech" industries such as computer hardware helped diversify their economies later in the decade.

Economy

The U.S. is first among nations in value of goods and services produced, followed by Japan. It has one of the world's highest per capita incomes and standards of living. Its large land area provides minerals needed for industry, and vast fertile areas produce more food than the nation needs. About one-third of the land is forested, supporting a wide variety of trees. Service industries are the largest economic sector, and include banking, insurance, wholesale and retail trade, and development. The main exports are automobiles, airplanes, computers, scientific measuring equipment, metals, paper, raw materials (such as metal ore), and chemicals (including raw plastics). Canada and Japan are the country's largest customers.

ALABAMA

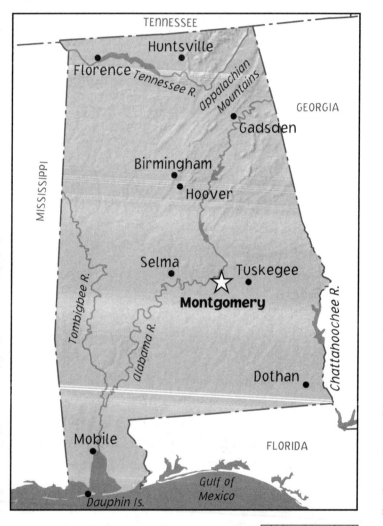

square miles: 50,744
population: 4,447,100
density: 88 people per square mile
capital: Montgomery
largest city: Birmingham
statehood: December 14, 1819 (22nd state)
nickname: Heart of Dixie
motto: We Dare Defend Our Rights
bird: yellowhammer
flower: camellia
tree: longleaf pine

Land

The southern end of the Appalachian Mountains shape northern Alabama into rugged, forested country. The mountains (the highest point being Cheaha Mountain at 2,407') fade away to the rich, dark soil that gives its name to the Black Belt, also known as the Cotton Belt because that plant was the main crop from the early 1800s until the 1940s. Tennessee, Tombigbee, and Alabama are the major rivers in the state. Alabama's southern edge is 53 miles of coastline along the deep indent of Mobile Bay and on the Gulf of Mexico.

Climate

Alabama has a hot climate, but in the dry northern mountains frost is common and snow sometimes falls. On the southern coastline, cooling sea breezes help on hot, humid summer days. July's average temperature is 80°F (27°C), and January's is 46°F (8°C). Yearly average rainfall is 5", slightly more on the coast. The growing season lasts 200 days in the north, 300 in the south.

People & History

For approximately 10,000 years, Native Americans occupied what became Alabama. People of the ancient Mississippian culture lived in towns and built many large burial and temple mounds that still can be seen today.

Their descendants—the Chickasaws, Cherokees, Muskogees (or Creeks), and Choctaws—lived throughout Alabama when the first Europeans arrived. They lived in villages, raising crops (such as corn), and tending orchards.

Spanish explorers first ventured into Mobile Bay and south Alabama in 1519. Conquistador Hernando de Soto, with a 500-man army, followed the Tennessee River into the north in 1540. De Soto's men fought resisting Choctaw warriors led by Chief Tuscaloosa, killing thousands in the bloodiest single battle between whites and Native Americans in the future U.S.

The Spanish and French explored, traded, and established settlements here over the next 250 years. Following the Louisiana Purchase, the United States gained control of this land in 1813. Alabama became a territory in 1817, achieving statehood two years later. Treaties with each of the four Indian nations, signed between 1830 and 1835, removed these tribes to Indian Territory (Oklahoma).

The first African Americans had been brought to the state as slaves in 1719. By 1860, slaves made up 45% of Alabama's 1 million residents. This reliance on slave labor served as a catalyst for Alabama to secede from the United States in 1861 and join the Confederate States of America. Montgomery, the state capital, was the Confederacy's first capital from April to July 1861.

Following the Civil War, federal troops remained stationed in Alabama during Reconstruction until 1876. By that time, Alabama law and social custom had created strict racial segregation. These Jim Crow laws lasted until the mid-1960s when the state became the scene of many landmark events in the civil rights movement. These include the Montgomery bus boycott of 1955–56, the 1964 deaths of four young African American girls in Birmingham from a bomb thrown into their Sunday school, and the 1965 march from Selma to Montgomery over laws preventing black citizens from voting.

Shortly afterward, the U.S. Congress passed the Voting Rights Act of 1965.

Industry

Cotton fueled the state's economy until the Civil War, even though manufacturing began before 1850. Textiles, cotton gins, and steel were being made here, but Alabama remained largely a farming state. In 1915, the one-crop cotton economy ended with a boll weevil infestation, although cotton is still an important part of the state's economy. Today mining, forestry, agriculture, and tourism are also major industries. Huntsville's Marshall Space Flight Center anchors the high-tech area. The Tennessee-Tombigbee Waterway also brings revenue to the state by offering a shorter shipping route than the Mississippi River from New Orleans northward.

Visiting Alabama

- Water sports attract visitors to the Gulf Coast and many public lakes, but snow skiing (on artificial snow) can be found at Cloudmont in the northeast corner.

- Golfing is available from the Gulf to northern Alabama at 18 courses of the Robert Trye Jones Golf Trail.

- Mobile's subtropical climate offers the Azalea Trail, 37 self-guided miles of azalea-blossom viewing in March and April through residential neighborhoods.

- Space travel—including a simulated shuttle trip and many rockets on display—is the theme at the U.S. Space and Rocket Center in Huntsville.

- Montgomery's Alabama Shakespeare Festival hosts a professional repertory company performing contemporary and classical plays from mid-November through August in two theaters on parklike grounds.

- In Theodore, Bellingrath Gardens and Home was the 900-acre estate of an early Coca-Cola bottler; 65 acres are gardens, and the house displays Dresden, Meissen, and Boehm porcelain collections amidst fine furnishings.

ALASKA

square miles: 571,951 (1st largest)
population: 626,932
density: 1 person per square mile
capital: Juneau
largest city: Anchorage
statehood: January 3, 1959 (49th state)
nickname: Land of the Midnight Sun;
The Last Frontier
motto: North to the Future
bird: willow ptarmigan
flower: forget-me-not
tree: Sitka spruce

Land

As the largest state, Alaska has 6,639 miles of coastline on the Pacific Ocean, Bering Sea, Arctic Ocean, and Beaufort Sea. Alaska's northern third is treeless Arctic Circle tundra; its Brooks Range is the northern end of the Rocky Mountains. The Alaska Peninsula and Aleutian Islands extend southwest from the Alaska Range in the state's center far into the Pacific. South of the Alaska Range, fertile inland plains are where most Alaskans live. The narrow, mountainous panhandle on the southeast borders

British Columbia. Alaska's highest point, Mt. McKinley in the Alaska Range, is also North America's highest at 20,320'.

Climate

Alaska has cool summers with very long days, cold winters with very short days. July average temperatures are 55°F (13°C), while January readings average 5°F (-13°C). Statewide average annual precipitation averages 55", mostly in snowfall.

People & History

Tlingit Indians of the southern coast traveled in oceangoing canoes as far south as Puget Sound for war and to capture slaves. Approximately 8,000 years ago, Eskimos and Aleuts began moving into the north, fishing and hunting the Arctic. The Tinneh, hunters and salmon fishers, lived in Yukon River valleys.

Sent by the Russian czar, Danish seaman Vitus Bering visited Alaska in 1728 and 1741; he discovered the sea otter population on his second trip. Russians began trading for furs, often holding Indian wives and children ransom for hides. Three Saints Bay, near Kodiak, was the first Russian settlement in 1784. From then until 1867, Russia controlled Alaska. Having made the sea otter nearly extinct and facing Crimean War expenses, it sold the land to the U.S. that year for two cents an acre.

In 1878, salmon canning along the southern coast began what became the world's largest salmon industry. Several gold discoveries, 1861–1903, added to the population, as did the Klondike gold rush into Canada, 1897–1900. Copper was found in Alaska in 1898.

During World War II, Japan invaded the Aleutian Islands, but was soon turned back. The U.S. built air bases and the Alaska Highway from Fairbanks into British Columbia, which is the main road to the lower 48 states.

Oil was discovered beneath Prudhoe Bay in the Arctic and on Kenai Peninsula in the 1950s; within twenty years oil became Alaska's main industry. After much controversy, the Alaska Pipeline to the southeastern port of Valdez was finished in 1977. In 1989, the world's largest oil spill occurred in Prince William Sound, severely damaging coastal ecology. Cleanup took three years and ecological recovery continues.

A major 1964 earthquake reshaped the panhandle, with land rising 16' at Cordova, destroying its harbor. Much of Anchorage was leveled, Valdez was devastated by a quake-caused tsunami ocean wave, and the coastline dropped 32' at Seward and Kodiak.

Industry

The oil industry dominates the state's economy. Alaska's fishing industry leads the states and related food processing makes more money than oil refining. In the southeast, pulp for paper is a major product. Government, including military bases, employs more people than any other economic segment.

Visiting Alaska

- Anchorage, on Cook Inlet below the Chugach Mountains, has symphony, opera, and light opera companies; Potter's Marsh, where visitors view waterfowl from boardwalks; the Alaska Zoo; and the Mount Alyeska ski resort. Alaska SeaLife Center in nearby Seward has an aquarium.

- Denali National Park surrounds Mt. McKinley. Visitors travel a two-lane highway in free buses that stop on demand. Backcountry camping is permitted, as are rock and ice climbing, river rafting, and air tours.

- Fairbanks, in Alaska's center, offers the Alaskaland cultural center with a reconstructed Indian village and Pioneer Museum. The University of Alaska Museum has exhibits on natural history and native cultures. A sternwheeler cruises the Chena River.

- Skagway's boardwalks and museum recall its days as a staging point for the Klondike gold rush, when the population was 20,000—twenty times that of today. White Pass and Yukon Route scenic railroad travels to Fraser, British Columbia.

- State-owned Alaska Marine Highway System's nine car ferries serve Pacific Ocean ports and connect to Prince Rupert, British Columbia, and to Bellingham, Washington. Cafeterias and dining rooms serve breakfast, lunch, and dinner.

- Juneau is the only U.S. state capital lacking highway access, but it has daily air- and seaplane service, and an ice-free harbor. Eskimo and Indian artifacts are exhibited at the Alaska Historical Library and Museum. The Juneau Icefield's 38 glaciers include Mendenhall, where walking and air tours are available.

ARIZONA

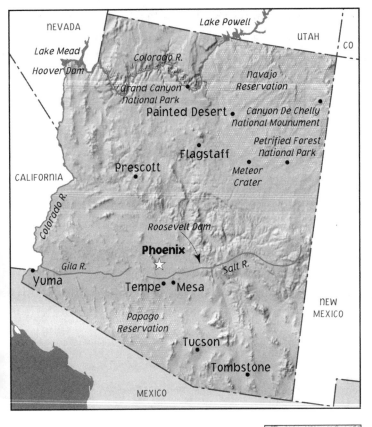

square miles: 113,635 (6th largest)
population: 5,130,632
density: 45 people per square mile
capital: Phoenix
largest city: Phoenix
statehood: February 14, 1912 (48th state)
nickname: Grand Canyon State
motto: *Ditat Deus* (God Enriches)
bird: cactus wren
flower: saguaro cactus flower
tree: paloverde

Land

Northeastern Arizona's Colorado plateau, plains, and mesas are cut by the Grand Canyon, where the Colorado River runs at 70' above sea level, but also include Humphreys Peak, the state's highest point at 12,633'. The rest of the state is open Basin and Range desert with occasional mountains.

Climate

Attracting snowbirds (human winter residents from the north), Arizona's average January temperature is 41°F (5°C). Desert summers see a July average temperature of 80°F (27°C). Only 1" of annual precipitation occurs, frequently in winter cloudbursts.

People & History

Around 1100, Hopi Indians built at Oraibi one of the oldest, still-used settlements in the U.S. The ancient Anasazi had moved away by the time Navajo and Apache Indians came here in the 15th or 16th century. Hohokam, who irrigated the southern desert, are ancestors of today's Pima and Tohono O'Odham Indians.

In 1539 and 1540, Spanish explorers sought rumored cities of gold here. Missionary Father Kino began exploring and founding missions in 1692, but the Indians resisted Spanish conquest. An adobe fort built at Tucson in 1776 protected Spanish soldiers. When Mexico won independence from Spain in 1821, future Arizona remained part of its New Mexico colony.

By the 1840s, Apaches had driven out most of the intruders. With the American victory in the Mexican War, which ended in 1848, Arizona to the Gila River came under U.S. control. The Gadsden Purchase, in 1853, gave the state its present southern border.

Arizona still was part of New Mexico when Confederate troops conquered it in 1862 during the Civil War. Union troops soon defeated them. The next year, both the Confederacy and the union recognized the Arizona Territory with its present boundaries.

Few settlers had braved the Indian-held lands, but an 1864 U.S. military expedition destroyed Navajo farms and homes, and the Navajo made the Long Walk to a reservation. Some Apaches also surrendered, but their warfare against settlers continued until 1886.

Arizona was a mining frontier with gold, silver, and copper discoveries. As early as 1867, farmers were irrigating the desert. In the 1870s and 1880s, major copper mines opened. World War I boosted cotton and copper production.

From 1911 through 1936, seven dams were built to store water for cities and agriculture. Cattle ranches, cotton plantations, and copper mines needed workers even during the Great Depression, and Arizona's population grew with U.S. immigrants seeking jobs during the 1930s. Further booms came with war-production increases and military airfields in the 1940s, as well as with returning veterans after the war and retirees later.

Today about one-fifth of the population is Hispanic and 6% is Native American. Arizona is one of the nation's fastest growing states, spurred in part by electronics firms moving here from the Northeast and California.

Industry

Manufacturing, the biggest economic sector, centers on Phoenix and Tucson, producing electronic equipment, guided missiles, and aircraft. Tourism is second, serving primarily winter visitors. Irrigated agriculture uses 85% of the state's water supply, with cotton the most important crop, followed by grains and alfalfa. Arizona produces 60% of U.S. copper today, and turquoise is mined.

Visiting Arizona

- Grand Canyon National Park protects the mile-deep, 277-mile-long canyon cut by the Colorado River. Driving, hiking, muleback, and helicopter are ways to explore.

- Phoenix's Heard Museum exhibits southwestern Indian art and crafts, and the Pioneer Living History Museum's 24 buildings reenact life from frontier to Victorian times, with live melodramas. Greyhound racing and casino gambling are available at Indian reservations. Golf courses and tennis courts are everywhere. The Desert Botanical Garden has 145 acres of Sonoran Desert flora.

- At Petrified Forest National Park visitors drive 28 miles through ancient trees, Indian petroglyphs, and the Painted Desert's pastel rock.

- Tombstone recalls the tough mining frontier of outlaws, gunfights, and saloons.

- Sedona is an art colony in the heart of the Red Rock Country's monoliths and mesas, with Rawhide simulating a frontier town.

- Canyon de Chelly National Monument on the Navajo Indian Reservation holds Anasazi cliff dwellings dating from 1100 A.D.

- Tucson includes the Children's Museum with hands-on exhibits, Reid Park Zoo, and the Arizona-Sonora Desert Museum with aquarium and desert creatures. There are also hiking or tram tours of Sabino Canyon's waterfalls and wildlife and the nearby Mission San Xavier del Bac dating from 1700. Saguaro National Park, on two units east and west of Tucson, holds forests of the tall (up to 40') raised-arm cactus that grows only in the Sonoran Desert.

- Manmade Lake Mead (shared with Nevada) and Lake Powell (reaching into Utah) are watersport meccas in the desert.

ARKANSAS

square miles: 52,068 (27th largest)
population: 2,673,400
density: 51 people per square mile
capital: Little Rock
largest city: Little Rock
statehood: June 15, 1836 (25th state)
nickname: Land of Opportunity
motto: *Regnat Populus* (The People Rule)
bird: mockingbird
flower: apple blossom
tree: pine

Land

Arkansas divides into two geographical zones along a diagonal line. The southeast half is lowland, part Mississippi flood plain and part western gulf coastal plain. The northwest half is highland, with the Arkansas River valley flowing through. North of the river, the Ozark Plateau has flat-topped ridges separated by steep valleys. Magazine Mountain, the state's highest point, rises to 2,753' from the Arkansas River valley. In the west's Oachita (wash-i-taw) Mountains are many hot springs.

Climate

Arkansas's humid subtropical weather includes 40" to 60" of precipitation a year, mainly in winter and spring. Little Rock, in the center, sees January average temperatures of 39°F (4°C) and July averages of 82°F (28°C). The northwestern mountains are slightly cooler and snow is possible. Tornadoes can occur in warm months.

People & History

French and Spanish explorers were the first Europeans to come into the area, where they found burial mounds from ancient Native Americans and met Caddo, Osage, and Quapaw Indian tribes. DeSoto's Spaniards crossed the territory of Arkansas in 1541–42. Frenchmen Marquette and Jolliet came down the Mississippi River (today's eastern border) to the mouth of the Arkansas River in 1673. Nine years later La Salle traveled the Mississippi, claiming its valley for France.

In 1686, a French trader built Arkansas Post on the Little Arkansas River for fur buying. It became an important stop for travelers between the Great Lakes and the Gulf of Mexico. Control of the land went from France to Spain (1762-1800) and back to France. But neither French nor Spanish settled here in any numbers.

Arkansas came under U.S. control with the Louisiana Purchase in 1803. The federal government forced Indian residents to move westward to Indian Territory (Oklahoma) over the next three decades. Settlers moved in—mainly English, Scottish and Scotch-Irish from Tennessee and Kentucky—and African-Americans were brought in as slaves for the state's southeastern plantations. The mountain people of the northwest became isolated, self-sufficient farmers.

Divided over secession, Arkansas did not join the Confederacy until the Union took Fort Sumter, South Carolina. Little Rock was captured by Union troops in 1863, beginning ten years of occupation. After war's end, former slaves became sharecroppers, required by law to live and work separately from whites. Arkansas continued as a poor and rural state until World War II, continuously losing residents who left in search of jobs.

Recent years have seen more residents living in cities than in the country; integration of schools, other public facilities, and state government; and economic diversification. Population, however, has not boomed because many people have moved away to seek jobs elsewhere.

Industry

Until the Great Depression, Arkansas was mainly agricultural, raising mostly cotton until the droughts of the 1930s. Today's agriculture includes rice (the state is the largest producer in the U.S.), chickens and turkeys, soybeans, and cotton. Manufacturing (including processing farm products) and tourism are the leading industries. Softwood and hardwood forests support the making of lumber, pulp, plywood, paper, and paper products. The production of electrical items and electronics is also important.

Visiting Arkansas

- In Hot Springs, Mid-America Museum's interactive exhibits present scientific principles. Window to the World Museum's guided tours cover 70 world cultures through artifact displays.

- Hot Springs National Park has 47 springs, bathhouses, and pools plus mountain roads, walking and horseback trails, and the Belle of Hot Springs sightseeing boat on Lake Hamilton.

- Eureka Springs, a former spa resort, today includes country music shows, historic home tours, The Great Passion Play (presented late April through October), and other Bible-themed attractions.

- Little Rock, Arkansas's largest city, holds the state capitol building (modeled on the U.S. Capitol), Little Rock Zoo, Arkansas Arts Center, and the Villa Marre Victorian mansion. The Arkansas Territorial Reconstruction tells of frontier life with five original buildings from the early 1800s, costumed interpreters, and visitor center exhibits.

- Mountain View's Ozark Folk Center presents regional music and dance, crafts, and oral history. Blanchard Springs Caverns can be viewed during one- or two-hour underground guided tours.

- Toltec Mounds State Park, near Scott, preserves earth mounds built for burials and as bases for homes and temples by ancient Indian people (not Mexico's Toltecs) that were abandoned around 1400.

- Buffalo National River in the Ozark Mountains offers hiking, camping, fishing, float trips, folk music, and craft demonstrations.

- Arkansas Post National Memorial, near Gillette, has a museum, visitor center, and townsite of the 1686 French fort that was the Mississippi Valley's first permanent European settlement.

CALIFORNIA

square miles: 155,959 (3rd largest)
population: 33,871,648
density: 217 people per square mile
capital: Sacramento
largest city: Los Angeles/Long Beach
statehood: September 9, 1850 (31st state)
nickname: Golden State
motto: *Eureka* (I Have Found It)
bird: California valley quail
flower: golden poppy
tree: California redwood

Land

California has 840 miles of Pacific Ocean coastline, but its lowest point is 86' below sea level in Death Valley. Coastal mountain ranges are 20 to 40 miles wide. Inland the Central Valley extends north to south, and on the east is the Sierra Nevada. In them, Mount Whitney is the state's highest point at 14,494'. Southeastern California holds the Mojave and Colorado deserts and the Salton Sea, which is formed by the flooding Colorado River.

Climate

California has many climate zones because of its size and range of elevations, but statewide January temperatures average 44°F (7°C), with July averaging a mild 75°F (24°C); humidity is generally low. Only 22" of precipitation falls annually, mostly in the north.

People & History

The main Indians living here were the Mohave (south), Maidu (north), and Yokuts (center). Spanish explorers sailed the coast in 1542, but Spain didn't colonize California until the 1770s. Priests built missions surrounded by ranches that Indians worked. California remained part of Mexico after the latter's independence from Spain in 1821; twenty years later the first American settlers arrived. Russian fur traders had built Fort Ross in 1812, but moved out by the 1840s. In 1846, unaware the Mexican War had begun, Americans captured Sonoma and declared a Republic of California.

Days before the war ended in 1848, gold was discovered at Sacramento, and thousands of prospectors rushed in. Statehood two years later brought homesteaders, and southern California's agricultural development began in the 1870s.

Many Chinese immigrants arrived beginning in the 1860s, where many worked in railroad construction. A publicity campaign in the 1880s brought thousands of Americans to southern California, and dependable sunlight attracted movie companies in the early 20th century.

Factories were built and shipyards grew when the nation entered World War I in 1917. Controlling the Colorado River's floods and channeling water for electrical power and irrigation was achieved with the huge Hoover Dam, completed in 1936.

World War II caused great industrial growth, and the building of military bases. Southern California boomed after the war, with freeways linking cities to expanding suburbs. The need for water in the dry south led to the contruction of a large canal and reservoir system completed in the 1970s. In 1963, California became the most populous state, and its population grew by a fourth between the 1980 and 1990 censuses.

Industry

California ranks first in the U.S. as a manufacturing and agricultural state. Aircraft, computers, missiles, and scientific instruments are important products. Farms specialize in crops, 20% are devoted to vegetables, 25% to fruits and nuts, and 30% to livestock. California leads the states in egg and milk production. The state holds a greater variety of minerals than any other. Service industries are the biggest economic sector, including tourism, scientific research, education, and government.

Visiting California

- Death Valley National Park, crossing into Nevada, holds brightly colored rocks, volcano craters, and sand dunes. Visitors can golf, swim, and ride horseback.

- Visitors and residents ride San Francisco's cable cars; Chinatown has shops and restaurants; San Francisco Maritime National Historical Park consists of ten sailing ships moored in the harbor; Fisherman's Wharf teems with street performers, shops, and restaurants, and it is the base for boat trips to Alcatraz Island prison. Museums include M. H. DeYoung Memorial Museum, Ansel Adams Center for Photography, and Wells Fargo Bank History Museum.

- Yosemite National Park has sequoia groves, waterfalls, granite cliffs and monoliths, a visitor center, and museums. Visitors can camp, hike, backpack, cross-country ski, and snowshoe.

- The twenty-one Spanish missions from the late 1700s and early 1800s are scattered along the Pacific coast.

- Monterey Bay Aquarium on Cannery Row, Monterey, holds multistory and smaller tanks with thousands of sea creatures. Converted canneries house shops and restaurants.

- Los Angeles, in addition to studio tours, has La Brea Tar Pits where prehistoric animals were fossilized, Old Spanish Plaza and Olvera Street, Chinatown, Little Tokyo, and many museums of art, popular culture, and natural history. Hollywood Bowl outdoor symphony concerts, plays on stages large and small, and professional sports are available.

- Lassen Volcanic National park, east of Redding, holds a volcano quiet since 1921, plus bubbling mudpots and steaming fumaroles.

- Disneyland, on 180 acres at Anaheim, opened in 1955 and is the sparkling granddaddy of theme parks.

- San Diego's zoo is one of the world's best; Old Town's Spanish buildings date from 1769; Junípero Museum tells of Spanish missions; Gaslamp Quarter National Historic District offers stores and restaurants; Old Globe Theatre presents modern and Shakespearean plays; Indian gambling casinos are near the city.

COLORADO

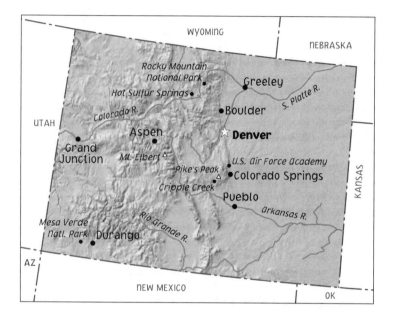

square miles: 103,718 (8th largest)
population: 4,301,261
density: 41 people per square mile
capital: Denver
largest city: Denver
statehood: August 1, 1876 (38th state)
nickname: Centennial State
motto: *Nil Sine Numine* (Nothing without Providence)
bird: lark bunting
flower: Rocky Mountain columbine
tree: Colorado blue spruce

Land

Colorado's land rises from the high plains at 3,350' on the Arkansas River through the hilly piedmont to the Rocky Mountains, which fill its western half. Its highest point, Mount Elbert at 14,433', is the highest mountain in the U.S. Rockies, and the whole state is the highest in the U.S. with average elevation of 6,790' above sea level. Major rivers flowing out of the mountains are the Arkansas, Colorado, Rio Grande, and South Platte.

Climate

Summer thunderstorms and winter blizzards, though severe, give Colorado little moisture, about 15" per year on the plains, but 42" in the mountains, mostly as snow. January temperatures average 28°F (-2°C) statewide, but July reaches averages of 74°F (23°C), with lower temperatures year-round in the mountains.

People & History

The Anasazi Indians moved into this land in the 12th century and left—for unknown reasons—in the 14th. They were followed four centuries later by nomadic Plains Indians—including the Arapaho, Cheyenne, and Ute—horseback hunters of buffalo.

Seeking imagined cities of gold, Spanish explorers reached Colorado in the 1840s. Spanish land grants extended to the Arkansas River in central Colorado, and military forts were built.

American exploration came after the Louisiana Purchase, beginning with Zebulon Pike's expedition in 1806. Fur traders and trappers, including famous mountain man Jim Bridger, followed, and trading posts were built.

When gold was discovered in central Colorado in 1859, many Americans rallied to the slogan "Pike's Peak or Bust," and the rush was on. By 1890, though, surface gold was gone and boomtowns abandoned.

To feed people in this dry land, pioneering farmers began to irrigate, and farming became more important to the state's future than the gold rush. Rich soil fed natural grasses that adapted to semiarid conditions and were excellent cattle and sheep feed. In the latter 19th century, natural resources of iron ore and coal led to steelmaking and other smokestack industries. Soon, about one-third of the land (the piedmont) was home to two-thirds of Coloradans, as it is today. Colorado's ethnic makeup is 87% white, 13% Hispanic, with smaller African American and Asian populations.

Industry

Tourism and service industries employ more than half of Coloradans; the state hosts the largest ski resorts in the U.S., but summer tourism is also very important. Federal employment, including military installations and national parks, is significant. Smokestack industries gave way to manufacturing of electronic equipment, computers, and telecommunication items. Livestock ranching and wheat farming still contribute greatly to the economy, but farmland has declined as irrigation water has been diverted to city use.

Visiting Colorado

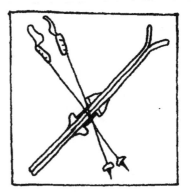

- Rocky Mountain National Park in north-central Colorado is craggy Continental Divide country, including 14,255' Longs Peak, wilderness areas, driving tours, and coyotes, black bears, and bighorn sheep. The gateway resort town Estes Park is the base for scenic drives; MacGregor Ranch Museum is an 1870s working ranch.

- Mesa Verde National Park in the southeast preserves Anasazi homes in cliffs near their farms. The park offers ranger-led tours up ladders to the homes, off-road bicycling, hiking, and camping.

- Florissant Fossil Beds National Monument west of Colorado Springs protects petrified trees, as well as insect, leaf, bird, and fish fossils. The visitor center has exhibits, and hiking and wildflower walks are offered.

- Denver's Larimer Square and LoDo (lower downtown) Historic District are among twenty such areas scattered among modern buildings, nightclubs, and restaurants. Molly Brown House remembers the Titanic survivor, who was also an early women's and children's rights activist. Visitors to the U.S. Mint watch coins being made. Red Rock natural ampitheater west of town is a scene for music performances. Professional sports include baseball, football, and hockey. The Denver Art Museum is near the Colorado State Capitol.

- Colorado Springs' city park, Garden of the Gods, holds unusual rock formations. Manitou Cliff Dwellings and Museum tell of the Anasazi. Museums range from the Pro-Rodeo Hall of Fame and the American Cowboy Museum to the Colorado Springs Fine Arts Museum and the World Figure Skating Hall of Fame and Museum. Nearby Pikes Peak has both a road and a cog railway to its 14,110' summit.

- Aspen is famous for its ski resorts, but it also has summer hiking, bicycling, fishing, and camping in national forests.

- Cripple Creek's streets were paved with gold—low-grade gold ore. More than $340 million in gold was produced here, and restored Victorian homes and businesses recall glory days. A narrow-gauge railroad tours four miles of area mines.

CONNECTICUT

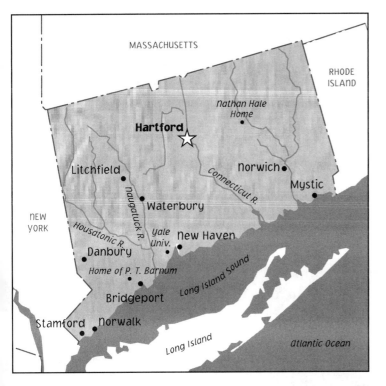

square miles: 4,845 (48th largest)
population: 3,405,565
density: 703 people per square mile
capital: Hartford
largest city: Bridgeport
statehood: January 9, 1788 (5th state)
nickname: Constitution State
motto: *Qui Transtulit Sustiner*
(He Who Transplanted Sustains)
bird: robin
flower: mountain laurel
tree: white oak

Land

From Connecticut's 90 miles of coastline on Long Island Sound (an inlet of the Atlantic Ocean), Mount Frissell rises to 2,380' in the rolling Berkshire Hills of the northwest. In the center of the state, the narrow, fertile Connecticut Valley has sandstone ridges above river valleys. The major rivers are the Connecticut and the Housatonic.

Climate

Although winter temperatures are relatively mild, Connecticut receives winter snows as part of its 36" to 48" annual precipitation. The average July temperature of 71°F (22°C) can be broken by heat waves, and the average January temperature of 26°F (-3°C) may dive in cold snaps.

People & History

In 1633, Dutch prospectors from New Amsterdam and British colonists from the Plymouth Colony built a post to trade with members of the sixteen Algonquin Indian tribes, but they built no settlements. The first white settlers, of the Massachusetts Bay Colony, moved into this land in 1633–35. In 1665, the English colony of Connecticut was created. Four years later the Fundamental Orders of Connecticut were written to guide government. It provided for a legislative assembly and an elected governor. This model was later used by the writers of the U.S. Constitution.

Residents supported the American Revolution, and 30,000 men volunteered for the Continental Army. Only a few skirmishes of the war occurred on Connecticut soil. At the U.S. Constitutional Convention in 1787, Connecticut members presented the Connecticut Compromise that outlined how U.S. representatives would be assigned to the states according to population.

In Colonial days, food grown here was exported to other colonies and also to West Indies sugarcane plantations. Connecticut depended on agriculture until the early 1800s when textile and other factories were built. By 1850, manufacturing employed more workers than farming. Eli Whitney, inventor of the cotton gin, sponsored machine-made (rather than handcrafted) parts for machinery and guns and founded a musket factory in Hamden. Plymouth clockmaker Eli Terry, in 1802, came up with the idea of interchangeable parts, which became the basis for all manufacturing. Samuel Colt, son of a textile manufacturer, invented the revolver and mass-produced it in Hartford beginning in 1836.

The state outlawed slavery in 1848. Besides supplying 55,000 men to the Union army during the Civil War, Connecticut supplied blankets, arms, and ammunition.

In the 1840s, Irish immigrants began to arrive and work in the state's factories. After the Civil War many French Canadians moved here. The late 19th century brought Italians, Poles, Hungarians, and Russians. Many African Americans came to Connecticut following World War II, and recent decades have seen Puerto Rican and Asian arrivals. Southwestern Connecticut is largely a bedroom community whose residents commute to New York City.

Industry

Connecticut's early gun factories led to its position as a major armaments supplier. Defense-spending cuts after the end of the Cold War (see Introduction) harmed these industries, but helicopters, tanks, aircraft engines, and submarines are also produced here. Factories producing cutlery and silverware, electrical items, chemicals, and textiles also are major employers. Hartford is an insurance-industry center, where many firms have headquarters. Oysters are the most important commercial seafood catch, and Connecticut leads the nation with its high supply. Casino gambling on Native American reservations has created tens of thousands of jobs in recent years, and tourism has increased. Significant in the state's agricultural sector are nursery plants, egg production, and feed for horses (the state has one of the highest per capita pleasure-horse populations in the U.S.).

Visiting Connecticut

- Essex Steam Train takes passengers on a 16-mile round trip from Essex Junction through the Connecticut River valley in early 20th-century style.

- At Groton, visitors may tour the USS *Nautilus*, the first nuclear submarine and the first submarine to pass under the North Pole's ice. Fort Griswold Battlefield State Park marks the site of a bloody 1781 Revolutionary War battle.

- Mystic Seaport at Mystic is a re-created 19th-century whaling village, with New England's last wooden whaling boat afloat in its harbor. Elsewhere in town, the Marinelife Aquarium exhibits 3,500 marine creatures and shows dolphins, sea lions, and whales at its indoor theater.

- Norwalk's Maritime Center exhibits the life of Long Island Sound in aquariums and has an IMAX theater; employees demonstrate boat-building.

- In Hartford, book lovers can visit the Mark Twain Mansion, where author Samuel Clemens lived in later life; it holds original furnishings and family belongings. A nearby cottage was the home of novelist Harriet Beecher Stowe. In West Hartford is the Noah Webster House, built in 1748, home to the dictionary creator. Colt Park was the estate of the revolver manufacturer.

- The Mashantucket Pequot Casino at Ledyard is the most profitable gaming house in the Western Hemisphere.

- New Haven, a major oil port on Long Island Sound, also offers the Yale Repertory Theatre and Long Wharf Theatre. Museums at Yale University include the Peabody Museum of Natural History and the Center for British Art.

DELAWARE

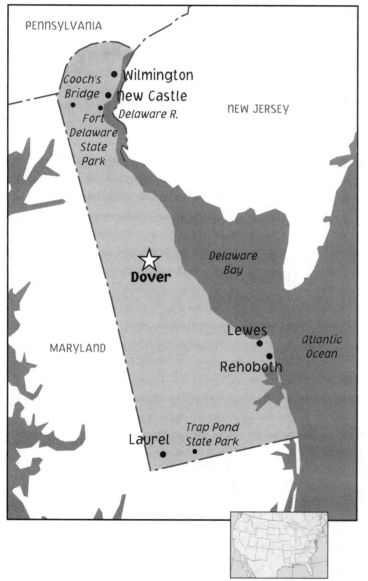

square miles: 1,954 (49th largest)
population: 783,600
population density: 401 people per square mile
capital: Dover
largest city: Wilmington
statehood: December 7, 1787 (1st)
nickname: The First State
motto: Liberty and Independence
bird: blue hen chicken
flower: peach blossom
tree: American holly

Land

Most of Delaware is a low plain with gentle hills rising to 442' in New Castle County in the north. After Florida, Delaware is the lowest-lying state of the nation. It has 28 miles of coastline on the Delaware Bay and the Atlantic Ocean and is located on the Delmarva Peninsula along with parts of Maryland and Virginia.

Climate

Delaware has a mild mid-Atlantic climate, with an average January temperature of

35°F (2°C) and July's average at 76°F (24°C). August is the rainiest month, receiving more than 5" of the annual 45" of precipitation.

People & History

In 1631, Delaware (Lenape) Indians killed the first European settlers here, Dutch from New Amsterdam (today's New York), the same year they arrived. Swedes made the first permanent settlement, Fort Christina at the site of today's Wilmington, seven years later. Legend says they built the first log cabins in America. In 1655, the Swedes were conquered by New Amsterdam's Dutch, who lost the colony to England nine years later. Although the Dutch briefly took it back in 1673, the colony was part of New York and then of Pennsylvania until 1704.

Delaware's delegate to the Continental Congress, Caesar Rodney, raced on horseback 80 miles to Philadelphia on July 1–2, 1776, to break a tie among Delaware members. His vote for independence allowed the Declaration of Independence to be issued on July 4. During the American Revolution, British troops heading for Philadelphia defeated local troops at Coochs Bridge (the state's only Revolutionary War battle), and the British navy harassed coastal towns.

In 1787, less than three months after the Constitution was written, Delaware became the first state to accept it and therefore was the first state in the Union.

Wilmington became the young nation's flour-milling center. In 1802, French immigrant Eleuthére Irenée du Pont built a gunpowder factory nearby, on Brandywine Creek. Industry flowered during the War of 1812 when British trade stopped on and off for three years. Many more factories opened in northern Delaware, today one of the nation's most industrialized areas.

Although Delaware allowed slavery, many slaves had been freed over the years after statehood. By 1860, only 1,800 slaves lived in the state, one-fifth of the number seventy years earlier. With trade ties to the North because of its factories, Delaware stayed in the Union during the Civil War. But since the Emancipation Proclamation did not apply to Union states, Delaware's slaves were not freed until the U.S. Constitution's 13th Amendment banned slavery in 1865.

Delaware increased construction of factories in the later 19th century. The growth of the railroad helped agriculture, moving produce to distant markets. In 1880, the Du Pont gunpowder factory began producing explosives and gradually expanded into various chemical products. Because of its dependence on manufacturing, the state was hit hard by the Great Depression of the 1930s, but it rebounded with World War II production.

The original British, Swedish, Dutch, and Finnish settlers were joined after the American Revolution by the French. Mid-19th century brought Catholic Irish and Germans; later immigrant groups included Ukrainians, Russians, Greeks, and more Scandinavians. Most immigrants settled in Wilmington, where most of the state's black population also lives.

Industry

Wilmington is Chemical Capital of the World, with one-third of its workers employed in chemical factories and research centers; E. I. Du Pont, Nemours & Company, Hercules Inc., and ICI (America) Inc. are headquartered here. With tax laws favorable to corporations, Delaware hosts many bank and insurance companies. State law allows businesses to incorporate here while operating elsewhere, which means a huge business community—on paper. Half of the state's land is farmed, with Delaware the largest U.S. producer of broiler chickens; leading crops are soybeans, corn, nursery products, and apples. Tourism is important on the southern Atlantic shore.

Visiting Delaware

• Henry Francis DuPont Winterthur Museum, near Wilmington, is an 1839 mansion housing a major collection of Early American furniture and art objects and sixty acres of gardens planted with native and exotic specimens.

• On nearby Brandywine Creek, the Hagley Museum and Eleutherian Mills are the original Du Pont home and gunpowder mill.

• The town of Rehoboth Beach was created by Methodists in 1872 for camp meetings. Today it is a major resort with a tiny permanent population and thousands of summer visitors who enjoy ocean sport-fishing, sailing, and swimming.

• Lewes has buildings and homes dating from its 18th-century seaport days, plus exhibits on Dutch settlers, local Indians, and ocean trade.

• New Castle's many historic buildings include Amstel House, dating from 1730 and housing colonial art and handcraft exhibits, and the 1732 Old Court House that served as the colonial and first state capitol.

• Visitors can take a boat to Fort Delaware on Pea Patch Island, offshore from Delaware City, which served as a Civil War prison.

FLORIDA

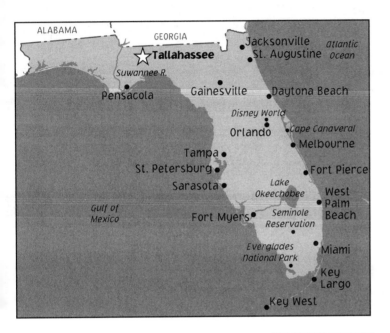

ALABAMA
GEORGIA
Jacksonville *Atlantic*
☆Tallahassee St. Augustine *Ocean*
Suwannee R.
Pensacola Gainesville Daytona Beach
Disney World
Orlando *Cape Canaveral*
 Melbourne
Tampa
St. Petersburg Fort Pierce
Sarasota *Lake*
Okeechobee West
Gulf of Palm
Mexico Fort Myers *Seminole* Beach
Reservation
Everglades Miami
National Park Key
Largo
Key West

square miles: 53,927 (22nd largest)
population: 15,928,378
density: 296 people per square mile
capital: Tallahassee
largest city: Jacksonville
statehood date: March 3, 1845 (27th state)
nickname: Sunshine State
motto: In God We Trust
bird: mockingbird
flower: orange blossom
tree: Sabal palmetto palm

Land

Most of Florida—southernmost of the contiguous 48 states—is under 100' in elevation, with 345' in Walton County the highest point. Inland are 4,424 square miles of freshwater—7% of Florida's area. North Florida is relatively hilly, central Florida has lakes (most of the state's 30,000 named lakes) and the main citrus belt, and south Florida is urban. Florida is primarily a peninsula between the Atlantic Ocean and the Gulf of Mexico; its 8,426 miles of coastline are second longest only to Alaska's. Off the south lie the Florida Keys, a 1,700-mile-long chain of coral-reef islands.

Climate

North of an imaginary line from Bradenton to Vero Beach, Florida is subtropical; south of there it is tropical. Occasional frosts occur as far south as Miami, but never in the Keys. January's average temperature is 59°F (15°C), July's 81°F (27°C). Average annual rainfall is from 40" (Key West) to 62" (West Palm Beach). Hurricanes can occur from June to November, mostly in September.

People & History

Native Americans came into Florida about 10,000 years ago, but European diseases and slavers had destroyed most of the Indian culture by 1750. Spanish explorers Ponce de Léon (in 1513 and 1521) and Cabeza de Vaca (1528) sought mythical riches. During the 16th century, France, Spain, and Britain fought over Florida. In 1565, Spanish soldiers massacred French Protestant settlers on the northeast coast, then founded St. Augustine nearby.

England traded Havana to Spain for Florida in 1763, and loyalist Florida was a base for raids on northern colonies during the American Revolution. After the war, England returned Florida to Spain, but later based troops in Pensacola during the War of 1812. United States troops captured Pensacola, which led to the U.S.'s gaining Florida at war's end.

The bloody Second Seminole War of 1835–42 resulted when the U.S. began to remove the Seminoles to Indian Territory. Most left for the west, but descendants of 300 who stayed currently live on three reservations in the Everglades.

Florida became a state in 1845, but joined the Confederacy during the Civil War. Battles were limited to Union captures of coastal cities. African slaves had arrived with Spanish explorers, and as the plantation economy developed in the 1800s, more were brought to Florida. By 1830, half the population was African American. Cubans moved here following an 1886 rebellion in their country and again for political reasons in the 1960s and 1980s. Greeks came to Tarpon Springs in the early 20th century, establishing the nation's major sponge industry. The Miami-Miami Beach area is home to a large Jewish community.

Industry

Before and after the Civil War, Florida was a sleepy agricultural state. In the 1880s, development financed by Northerners began: mining of phosphate (Florida's most important mineral, used in fertilizer), and tourism.

Tourism is by far Florida's most important industry, followed by manufacturing (especially food processing) and agriculture. Florida produces three-fourths of the nation's citrus fruit and is second only to California in vegetables. Breeding thoroughbred horses is significant. Military bases are economically important, and Brevard County is home to Kennedy Space Center.

Visiting Florida

- Everglades National Park extends 40 miles inland from the southeastern tip of Florida and is home to tropical plants and trees, manatees, crocodiles, and snakes. Boating, canoeing, natural trails, and camping are offered.

- Walt Disney World Resort near Orlando covers 28,000 acres and includes theme parks, golf courses, camping, luxury spas, an animal park, Disney-MGM Studios, and live and animated entertainment.

- St. Augustine, founded in 1565, is the oldest European-founded settlement in the U.S.

- Edward Ball Wakulla Springs State Park, southwest of Tallahassee, includes the spring Ponce de Léon believed to be the Fountain of Youth, narrated river and glass-bottom boat tours, birding, hiking, and bicycling.

- Bus tours take visitors through the Kennedy Space Center, east of Orlando. Site includes IMAX theaters, the U.S. Astronaut Hall of Fame, and the Air Force Space Museum.

- On the Gulf of Mexico and Tampa Bay, St. Petersburg and Tampa are home to Busch Gardens Tampa Bay, the Florida Seaquarium, Museum of Science and Industry (with a simulated hurricane), and the Salvador Dali Museum.

- Cypress Gardens at Winter Haven, opened in 1936, is Florida's first theme park and has 208 acres of botanical gardens, free-flying butterflies in a conservatory, and a water-ski show.

- The Miami-Miami Bay area offers pro baseball, basketball, and football; symphony, ballet, and opera; Biscayne National Park wilderness with coral reefs; Italian Renaissance villa Vizcaya Museum and Gardens furnished in ornate European styles; Parrot Jungle and Gardens where exotic birds perform; a cageless Metrozoo; the Miami Seaquarium; Little Havana; luxury hotels; and a beach boardwalk.

GEORGIA

TENNESSEE   nc

Appalachian Mts.
- Chickamauga Battlefield
- Rome
- Athens
- Atlanta ☆
- △ Stone Mtn.
- Augusta
- Warm Springs
- Macon
- Little White House
- Warner Robins
- Columbus
- Albany
- Savannah
- Brunswick
- Jekyll Is.
- Valdosta

Hartwell Lake
Clark Hill Lake
SOUTH CAROLINA
Savannah R.
Altamaha R.
Chattahoochee R.
Flint R.
ALABAMA
Okefenokee Swamp
St. Mary's R.
Lake Seminole
FLORIDA
Atlantic Ocean

square miles: 57,906 (21st largest)
population: 8,186,453
density: 141 people per square mile
capital: Atlanta
largest city: Atlanta
statehood: January 2, 1788 (4th state)
nickname: The Empire State of the South
motto: Wisdom, Justice, Moderation
bird: brown thrasher
flower: Cherokee rose
tree: live oak

Land

The Blue Ridge portion of the Appalachian Mountains extends into north-central Georgia, with Brasstown Bald (4,784') their highest peak. The Piedmont region to the south holds abruptly rising mountains like Kennesaw and Stone. The land drops in the Fall Line area to sandy hills, then to the rolling terrain of coastal plain and flat pine barrens along 100 miles of Atlantic coast. In the extreme southeast, Okefenokee Swamp extends from Florida.

Climate

Georgia's climate is mild and moist, with 50" of rain spread throughout the year in the north, and 44" in the east central. The coast and south receive most of their rain during summer. The average January temperature is 47° F (8°C), and July, 80° F (27° C).

People & History

By the year that Spain's Hernando de Soto arrived here, in 1540, the ancient mound-building cultures were giving way to those of the Cherokees and Creeks. Spain sent missionaries to offshore islands within 30 years.

Britain wanted to block Spain and granted its last charter for a colony to James Oglethorpe in 1732. He intended to make a home for free-thinkers and other social outcasts from various nations. Along with English and Scots, he welcomed Portuguese Jews and German and Swiss Protestants. He banned slavery and the rum trade. But fighting off Spanish invaders hurt development and Oglethorpe put control of the colony back in the hands of Parliament.

Settlers then poured in from northern colonies, and Georgia prospered. Many residents didn't want to join the American Revolution, but relented as troops from other colonies attacked Georgia.

In 1838, the entire Cherokee nation (17,000 men, women, and children) was removed from northwestern Georgia. Their journey—called the Trail of Tears because it was marked by over 4,000 deaths—was a dark chapter in Georgia's history.

Before the Civil War, residents divided between proslavery planters and anti-slavery farmers and frontiersmen. Georgia seceded from the Union in January 1861 and suffered great casualties during the war. By its end, most of Atlanta, a Confederate supply and railroad hub, had been burned, and most of Georgia's wealth and young men were gone.

After the war, Georgia recovered better than many other Confederate states. The central cotton-producing country suffered the most, but elsewhere in the state new factories made cotton textiles and processed white clay for ceramics, and forests were turned into lumber. Coca-Cola, made in Atlanta, became a major export. Atlanta, the state's largest city, became important for the entire South: a center for finance, the arts, and higher education. Georgia continued racial segregation after the Civil War, but the civil rights movement that ended it during the 1960s used mostly purchasing boycotts, lawsuits, sit-ins, and other nonviolent demonstrations.

Industry

Forests cover two-thirds of the state, carefully replanted as trees are cut. Wood and paper products are important along with giant poultry businesses, where chickens are raised entirely indoors. Textiles and Coca-Cola support large factories. This state is first in the U.S. in growing peanuts and pecans; other crops include sugarcane, peaches, watermelon, okra, and pimento peppers. Marble and granite are quarried and industrial minerals produced.

Visiting Georgia

- Chickamauga and Chattanooga National Military Park interprets the Confederate victory at Chickamauga Creek, September 1863, and losses at Chattanooga and Missionary Ridge, November 1863.

- Fort Pulaski National Monument, west of Savannah, shows life at a military post of the Civil War era.

- Savannah's history includes British Colonial life, the Revolutionary War, "King Cotton," and the Civil War. A self-guided tour winds among many buildings dating from the 1700s.

- Ocmulgee National Monument in Macon protects Indian mounds and artifacts from prehistoric days to the time of the Creek Indians.

- Owens-Thomas House and Museum, Savannah, is a furnished, Regency-style English villa dating from 1819.

- Jekyll Island and Sea Islands are beach resorts with night-time entertainment, sea-fishing, and wild-life refuges.

- Tifton's Georgia Agrirama takes visitors back to the 1890s in an outdoor, living history museum that includes a town, sawmill complex, and farm communities.

- Hofwyl-Broadfield Plantation near Brunswick was a slave-operated rice plantation before and after the Civil War.

- Atlanta offers city fun and education: the state capitol; professional baseball, football, and basketball; Atlanta Historical Society's museum with exhibits on the Civil War, African-American history, folk art, the film *Gone with the Wind*; automobile racing; the Martin Luther King Jr. National Historic Site; CNN studio tours; canoeing and rafting on the Chattahoochee River.

- Georgia's Stone Mountain Park, outside Atlanta, displays gigantic mountainside carvings of Jefferson Davis, Stonewall Jackson, and Robert E. Lee.

- Andersonville National Historic Site, the location of an infamous prisoner of war camp for Union soldiers, serves as a monument to any American ever held as a prisoner of war.

HAWAII

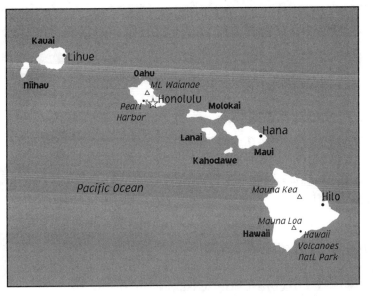

Kauai
• Lihue
Niihau
Oahu
Mt. Waianae
△ • Honolulu
☆
Pearl
Harbor
Molokai
Lanai
• Hana
Maui
Kahodawe
Pacific Ocean
Mauna Kea
△
• Hilo
Mauna Loa
△
Hawaii
• Hawaii
Volcanoes
Natl. Park

square miles: 6,423 (47th largest)
population: 1,211,537
density: 189 people per square mile
capital: Honolulu
largest city: Honolulu
statehood: August 21, 1959 (50th state)
nickname: The Aloha State
motto: *Ua Mau Ke Ea O Ka Aina I Ka Pono*
(The Life of the Land Is Perpetuated in
Righteousness)
bird: Nene
flower: yellow hibiscus
tree: Kukui

Land

The southernmost state includes 132
islands with 750 miles of Pacific Ocean
coastline, situated 2,400 miles from the
mainland U.S. Most Hawaiians live on
the eight main, southeast, islands—Kaui,
Molokai, Lanai, Kahoolawe, Maui,
Hawaii (the Big Island) and Oahu—and
privately owned Niihau. These islands
were created by volcanoes, and their two
active volcanoes today are Mauna Loa
and Kilauea, on Hawaii. Dormant vol-

45

cano Mauna Kea is also on Hawaii, and it is the state's highest point, reaching 13,796'. In the middle of the volcanic chain are rock islets; coral and sand islands are to the northwest.

Climate

Hawaii has mild temperatures year-round, with only 8° between January's average 72°F (22°C) and July's 80°F (27°C). Summers are cooled by constant northwest trade winds off the Pacific. Precipitation varies widely with location, with more falling on the islands' windward sides than on their leeward (away from the wind) sides, and more at higher elevations. Only 23" a year falls on Honolulu at sea level on Oahu, but Kauai's Mount Waialeale is one of the earth's wettest places with 480" per year. Snow falls on all mountains above 9,000'.

People & History

Polynesians who probably arrived here by canoe around 400 A.D., developed a culture ruled by kings and priests. Items for daily life were made from materials such as stone, bone, shells, and wood. However, metal-working and pottery-making had not developed when Europeans arrived.

Possibly not the first European to see Hawaii, Britain's Captain James Cook reached Kauai early in 1778, and he was killed by natives when he returned in 1779. Captain George Vancouver introduced livestock in 1794; in the early 1800s, crews of whaling ships spent their winters in the islands.

In 1810, King Kamehameha I united all the islands, beginning 83 years of monarchy. Christian missionaries began to introduce European and American culture in 1820, but Hawaiian leaders began to work hard to preserve their own culture. King Kamehameha III initiated constitutional government in 1839. When Queen Liliuokalani ascended the throne in 1891 and tried to change the constitution to increase royal power, she was overthrown by a coalition of Hawaiians and American business interests. They established the Republic of Hawaii in 1893, which lasted seven years.

Despite promising to protect Hawaii's independence, Britain, France, and the U.S. all wanted to control these islands that were strategically located in the Pacific. In 1898, the U.S. annexed Hawaii, making it a territory in 1900.

The population has grown diverse. Beginning in the 1850s, sugarcane plantations attracted Chinese, Japanese, and Polynesian workers. In the early 20th century, immigrants from Asia, the Philippines, Europe, and America swelled the population, outnumbering native Hawaiians.

Industry

Millions of visitors each year make tourism the most important private industry in Hawaii.

Federal military bases and other government agencies are important public employers. Hawaii leads the states in producing pineapples and sugarcane on huge plantations, but agriculture today is a small portion of its economy. Cattle, hogs, chickens, avocados, macadamia nuts, and ornamental shrubs are raised. Manufacturing—including canned pineapple, clay and glass products, and ships—accounts for about 5% of the state's income and employment.

Visiting Hawaii

- Hawaii Volcanoes National Park, on Hawaii, surrounds the island's two active volcanoes. Visitors can choose easy or demanding hikes, an 11-mile scenic drive, and explore desert, rain forest, and arctic tundra ecosystems.

- Maui holds Lahaina, the royal capital city, Carthaginian II Floating Museum's whaling exhibits; Whaler's Village; and half-hour rides on a replica narrow-gauge sugercane railroad.

- Honolulu offers Iolani Palace, home to the last kings and queens; Aloha Tower shops; the Hawaii Maritime Museum, Waikiki Beach high-rise hotels as well as nightclubs; the Pacific Aerospace Museum; hula shows; the Bernice P. Bishop Museum; the Planetarium's Hawaiian cultural exhibits; the Honolulu Zoo; and the scenic Top-Tantalus drive.

- Pearl Harbor near Honolulu includes the USS *Arizona* Memorial Visitor Center (with a short film on the December 7, 1941, Japanese attack that drew the U.S. into World War II) and the shrine built on the water above the Arizona; which was sunk that day.

- Makapuu Point's Sea Life Museum on Oahu includes shows; a gigantic Hawaiian reef tank; and the Pacific Whaling Museum.

- Molokai is home to Father Damien's settlement for lepers; reefs where visitors snorkel and scuba dive; and the world's highest sea cliffs.

IDAHO

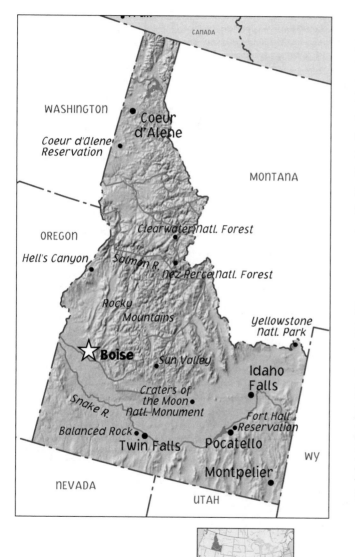

square miles: 82,747 (13th largest)
population: 1,293,953
density: 16 people per square mile
capital: Boise
largest city: Boise
statehood: July 3, 1890 (43rd state)
nickname: The Gem State
motto: *Esto Perpetua* (It Is Forever)
bird: mountain bluebird
flower: syringa
tree: Western pine

Land

Southern Idaho is the sagebrush Columbia Plateau along the Snake River, which flows west. The state's lowest elevation is 710' on the river; mile-deep Hells Canyon on the Snake is North America's deepest gorge. Most of the rest of Idaho is rugged Rocky Mountains, with its highest point at 12,662' Borah Peak. The fertile Palouse hills in the northwest are also part of the Columbia Plateau.

Climate

Idaho's mountains cause cool summers, the July average temperature is 67°F (19°C); and cold winters, the January average temperature is 23°F (-5°C). Much of the 19" of annual precipitation is in the form of snow.

People & History

Nomadic Kutenai, Kalispel, Coeur d'Alene, and Nez Perce Indians lived in small villages in northern Idaho, and Shoshone and Paiute live in the south. The Lewis and Clark Expedition in 1805 included the first whites known to enter this land. Only four years later, the first fur-trading post was built on Pend Oreille Lake.

Gold discovered in 1860 brought prospectors to the mountains. Most miners were former Confederates who arrived during and after the Civil War. Also in 1860, farmers, many of them Mormons from neighboring Utah, moved into southern Idaho, and a strong regionalism based on lifestyles developed.

Major lead and silver discoveries in 1880 and 1884 caused the creation of underground mines. In 1892 and 1899, labor troubles between union and non-union miners and between management and labor led to violence. At the same time, cattle and sheep ranchers fought over the limited rangelands.

In the early 1900s, the timber industry rose to prominence, along with agriculture. Idaho's economy was severely harmed when farm prices dropped in the 1920s and when the Great Depression followed in the 1930s. Federal reclamation projects and military-base building helped the state recover. Industry diversified in the 1940s, lessening dependence on food and mineral prices. In 1949, Idaho's largest federal installation, the Nuclear Reactor Testing Station, was built.

Idaho residents work to balance economic growth with maintaining their rich natural environment.

Industry

Agriculture is Idaho's most important industry, led by potatoes, wheat, and sugar beets. Manufacturing is a close second, including food processing and wood products, mobile homes, and electronic and construction equipment. Idaho mining produces more silver than any other state; other important minerals are lead, gold, and phosphate. Tourism to state and national parks brings in hundreds of thousands of dollars a year.

Visiting Idaho

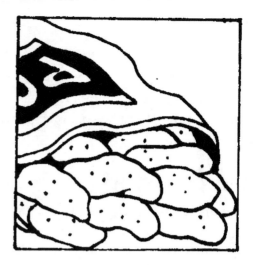

- Famous for its bright sun and deep snow, Sun Valley is a four-season resort founded by the Union Pacific Railroad in 1936.

- Boise is an outdoor city in summer and winter, offering water sports on the Boise River and a variety of recreational activities in Boise River greenbelt lands. Underwater windows at the MK Nature Center allow unique views of the river's natural life. The Julia Davis Park includes Boise's Art Museum, the Idaho State Historical Museum, Zoo Boise, the Discovery Center, and a rose garden. Payette River Scenic Byway leads to hiking, fishing, and several types of boating.

- Nez Perce National Historical Park extends through Idaho, with a visitor center at Spalding and separate sites telling the story of the Nez Perce Indians' 1877 flight away from reservation life. The Lewis and Clark Expedition also passed this way in 1805 and again in 1806.

- At Twin Falls, 212' Shoshone Falls of the Snake River is higher than Niagara Falls, and the city borders rugged Snake River Canyon.

- Silver Valley's entire town of Wallace is a National Register of Historic Places' site, with museums and a mine tour that tells about the silver boom. Shoshone County Mining and Smelting Museum is at Kellogg. Silver Mountain Resort has a 3.1-mile gondola ride.

- Craters of the Moon National Monument northwest of Twin Falls shows the marks of volcanic activity two millennia ago; a loop drive wanders among lava cones, towering granite cliffs, and natural bridges.

- In Hells Canyon National Recreation Area, visitors can raft or jet-boat the Snake River, hike trails, fish, camp, or hunt.

- Coeur d'Alene, on the lake of the same name, is a city of resorts, golf courses, antique shops, restaurants, the Silverwood Theme Park, excursion boats, and all kinds of water sports.

ILLINOIS

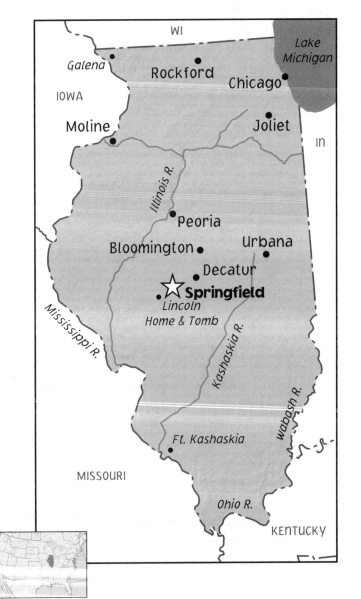

square miles: 55,584 (24th largest)
population: 12,419,293
density: 223 people per square mile
capital: Springfield
largest city: Chicago
statehood: December 3, 1818 (21st state)
nickname: Prairie State, Land of Lincoln
motto: State Sovereignty, National Union
bird: cardinal
flower: native violet
tree: white oak

Land

Most of Illinois is flat prairie, but rolling hills in the northwest rise to the state's highest point, Charles Mound, at 1,235'. The lowest point is 279' on the Mississippi River, Illinois's western boundary. Lake Michigan, one of the Great Lakes, forms the northwestern edge with over 60 miles of shore. In southern Illinois, the crested Shawnee Hills are low, forested, and separated by plateaus. The Illinois River is the major in-state river, and the Wabash River marks the southern border.

Climate

Illinois has cold, snowy winters and hot, humid summers. The average January temperature is 26°F (-3°C), with July's average at 76°F (24°C). Annually, the state receives 38" of precipitation, with averages higher in the south.

People & History

The largest ancient Indian town north of Mexico, Cahokia, located on the Mississippi, was home to 10,000 people from 850 to 1150. Indians of the historical period included Illinois, Kickapoo, Sac, Fox, Potawatomi, Ottawa, and Chippewa.

French explorers Louis Jolliet and Jacques Marquette reached this area in 1673 while navigating the Mississippi, and René Cavalier Robert Sieur de la Salle built the first French fort near Peoria in 1680. After Britain won the land from France in 1763, it tried to ban further white settlement so that it had a smaller area to defend.

During the Revolutionary War, Americans captured Fort Kaskaskia from the British in 1778 and established Illinois as a Virginia county. It became part of the Indiana Territory in 1800, and its own territory nine years later.

Indians trying to prevent white settlement were defeated in the Black Hawk War (1832). Population boomed with the National Road and increased canal-building in the 1840s; plus the arrival of railroads in the 1850s. Chicago soon became—and remains—the nation's railroad hub.

The people of northern and southern Illinois were divided on the question of slavery, and some Southern residents tried to break off their region to join the Confederacy. However, the state remained whole and in the Union during the Civil War.

Waves of European immigrants and freed African Americans moved into Illinois after the war, working in factories, railyards, and meat-packing plants. When Chicago was destroyed by fire in 1871, it was promptly rebuilt. Early attempts to organize labor unions were significant here, including the Pullman (railroad) strike in 1894 and the Haymarket Square Riot of 1886 in Chicago. The latter began after a bomb exploded at a workers' rally, leaving eight people (seven of them policemen) dead.

Chicago is justly famous for its gangsters of the Prohibition era, but Downstate (the rest of Illinois) had its share of dark activities, including a long-running family feud among Williamson County residents and the Ku Klux Klan in the 1920s.

Industry

The third-largest manufacturing state, Illinois produces nonelectrical machinery, food items, electrical equipment, chemicals, and petroleum. The Midwest Stock Exchange, Chicago Mercantile

Exchange, and Chicago Board of Trade (which sets world agricultural prices) all are located here. O'Hare Airport in Chicago is one of the world's busiest, and freight and passenger trains use Chicago's vast railyards. The Illinois Waterway allows ships to travel from Lake Michigan to the Mississippi River. Corn and soybeans, cattle and hogs are the greatest agricultural products.

Visiting Illinois

- The "Land of Lincoln" includes Lincoln's New Salem State Historic Site near Springfield, which reproduces the town where the then-future president lived 1831–37; in Springfield is the white frame Lincoln family home of 1844–61; and the State House at Vandalia was the capitol when Lincoln served as state representative.

- At Galena, the home of another president, Ulysses S. Grant, appears as it did when Grant and his family moved there in 1860.

- Oak Park includes buildings designed by Frank Lloyd Wright, who lived here.

- Chicago is the nation's third-largest city. Visitors enjoy Old Town's restored homes, nightclubs, and shops. The Chicago Zoological Park, with moats rather than bars between visitors and animals, is considered one of the nation's best zoos. The tallest U.S. building is 110-story Sears Tower. At the intersection of Michigan and Chicago avenues stands the Old Water Tower, which survived the Great Chicago Fire. Symphony, opera, and theater are available. French impressionist art is exhibited at the Art Institute of Chicago. The Field Museum of Natural History is across from the John G. Shedd Aquarium. Professional sports include two baseball teams, and football, basketball, and hockey teams.

- The leader of the Mississippian Indians and his family lived in the capital, Cahokia, atop ten-story Monks Mound, the largest Indian earthwork in the U.S. today, which is near Collinsville.

- A Swedish religious commune, Bishop Hill near Galva, was founded in 1846. Original buildings remain, and a collection of American primitive paintings is exhibited.

- Shawnee National Forest, the state's only federal land, sprawls across nine southern counties, holding forest trails, campsites, fishing spots, and scenic drives.

INDIANA

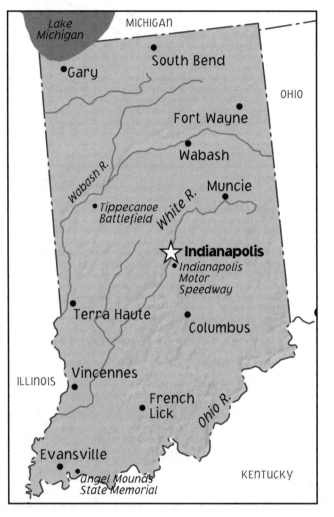

square miles: 35,867 (38th largest)
population: 6,080,485
density: 170 people per square mile
capital: Indianapolis
largest city: Indianapolis
statehood: December 11, 1816 (19th state)
nickname: The Hoosier State
motto: The Crossroads of America
bird: cardinal
flower: peony
tree: tulip tree

Land

Indiana stretches from the sand dunes on its 41 miles of Lake Michigan shore on the north to the banks of the Ohio River on the south. South of the Great Lakes Plains lie the Till Plains, holding richly fertile soil. The Southern Hills and Lowlands include steep hills called knobs. Indiana's highest point, 1,257', is in Wayne County near the Ohio state line, and its lowest where the Wabash River flows into the Ohio River, 320'.

Climate

Humid, hot summers and cold winters shape Indiana's climate. The state receives 40" of precipitation per year, with July average temperatures of 75°F (24°C), and January average temperatures of 28°F (-2°C). Summers are slightly cooler and winters slightly warmer—but with more snowfall—along Lake Michigan.

People & History

Ancient mound builders lived on the Ohio River near Evansville, but when the first Europeans reached this land, residents included mainly Miami, Potawatomi, Shawnee, and Delaware Indians. In 1769, first René Cavalier Robert Sieur de la Salle, a Frenchman, arrived from the north on the St. Joseph River. French traders and forts followed, with Fort Vincennes in 1732 among the first permanent white settlements west of the Appalachians. Meanwhile, English settlers moving from eastern colonies settled along the Ohio and Wabash rivers.

Indiana came under British control in 1763. During the Revolutionary War, the Continental Army captured Indiana from Britain (1779).

Indians continued to fight for their lands until their defeat in 1794 at the Battle of Fallen Timbers in Ohio, and then again from about 1800 until the Battle of Tippecanoe in 1811. From 1820 to 1840, all major tribes moved away.

Because it was part of the Northwest Territory, where slavery was banned, Indiana entered the Union as a free state. In 1800, though, there were 175 existing slaves who were not freed upon statehood. African Americans were prohibited from moving into Indiana by state law until 1866.

Indiana became an important agricultural state in the first half of the 19th century. The Civil War brought industry, especially in the northern part of the state. Beginning in 1906, Gary became a steel-making center, using iron ore from the Mesabi Range in Minnesota.

The original British colonial settlers were joined by Irish and German immigrants in the 19th century. Factory employment after World War I attracted other Europeans, and African Americans.

Industry

Seventy percent of Indiana land is agricultural but contributes only 1% of the gross state product. Corn and soybeans are the most important crops. Manufacturing, mainly in the northeast around Gary but also in Indianapolis, accounts for one-third of the state's economy and employs one-fourth of its workers. Products include iron, steel, petroleum products, and automobile parts. Services, including personal and business, finance and real estate, make up more than one-fourth of the economy.

Visiting Indiana

- Indianapolis's greatest annual event is the Indianapolis 500 automobile race—and related festivities—on Memorial Day weekend. In the 1990s, the Brickyard 400 stock-car race was added in August, and planning began for a Formula One race. The Indianapolis Motor Speedway and Museum, with cars and other exhibits, is open year-round. The five floors of the Children's Museum house dinosaurs and interactive exhibits. White River State Park includes a zoo. Museums include the Eiteljorg Museum of American Indian and Western Art, the Indianapolis Museum of Art, and the Indiana State Museum with natural history and paintings by Indiana artists. Professional teams play football and basketball.

- Indiana Dunes National Lakeshore protects Lake Michigan beaches for public use. The dunes—reaching to 180' above the water—support an unusual mixture of temperate-climate and desert plants.

- Vincennes's George Rogers Clark National Historical Park covers the rebelling American colonies' capture of this French-founded city and General Clark's career. The wooden capitol of the Indiana Territory and Grouseland—home of territorial governor (later U.S. president) William Henry Harrison—recall the early 19th century.

- Columbus features a historic downtown and restored Victorian architecture along with 20th century structures by modern architects including Eero Saarinen.

- Connor Prairie Settlement at Nobelville allows visitors to view the life of a prairie settlement of 1836, with costumed interpreters and audience participation.

- Amish country in northern Indiana's Elkhart and LaGrange counties is where these "plain people" live without using electricity or gasoline because of religious beliefs. They share their German-heritage foods, crafts, and horse-drawn buggy rides with visitors.

IOWA

square miles: 55,869 (25th largest)
population: 2,926,324
density: 52 people per square mile
capital: Des Moines
largest city: Des Moines
statehood: December 28, 1846 (29th state)
nickname: Hawkeye State
motto: Our Liberties We Prize and
Our Rights We Will Maintain
bird: eastern goldfinch
flower: wild rose
tree: oak

Land

Iowa's rolling hills lie between the Mississippi River on the east and the Missouri on the west. Its lowest point is 480' on the Mississippi, and its highest is 1,670' in Osceola County in the northwest. Glaciers scraped the land, having least effect in the northeast, where rugged hills and cliffs remain. Along the Missouri, ancient piles of windblown soil formed bluffs above the river.

Climate

Hot, rainy summers and cold, snowy winters mark Iowa. January's average temperature is 19°F (-7°C), with July's average rising to 75°F (24°C). Annual average precipitation is 32", nearly three fourths falling as rain from April through September.

People & History

Paleo-Indian hunters roamed Iowa beginning 13,000 years ago, followed by Woodland village-builders and Adena mound builders along the Mississippi River until about 800 A.D. When Frenchmen Louis Jolliet and Jacques Marquette explored the Mississippi in 1763, Iowa Indians and Santee and Yankton Sioux lived here. They were driven out by the Wisconsin and Illinois tribes, themselves pushed west by white settlers.

Iowa's first settlement came when Julien Dubuque started a lead mine in 1788, leading to the town of Dubuque 45 years later. Control of Iowa passed from France to the U.S. with the Louisiana Purchase of 1803 (see Introduction). Settlers from other states began to arrive in the 1830s and 1840s, meeting Indian resistance until the Black Hawk War of 1832. Later immigrants arrived mainly from Sweden, Holland, and Germany.

Iowa became an agricultural state, with specialized farms that produced for sale rather than for family use. Riverboats, mainly in the 1850s to 1870s, created freighting centers on the Mississippi and Missouri, loading food for market and supplies for the northwest frontier.

Iowa entered the Union as a free state in 1846, counter-balancing the entry of slave-state Florida in 1845. Iowans hosted Underground Railroad stops, helping slaves escape to freedom. A greater percentage of Iowans fought in the Civil War than from any other Northern state. Following the war, Iowa opened voting to African Americans and integrated schools.

By the 1870s, four railroad lines served farmers; today twenty freight railroads run. Most Iowans lived in rural settings until 1960, forty years after most U.S. residents lived in cities.

The rivers that make Iowa's soil rich also cause damage. Despite dams for flood control, power, and transport, rivers rage as the Mississippi did in 1993, causing $2 billion in damage.

Industry

Agriculture dominates Iowa's economy, with more than 90% of land farmed. Iowa produces 7% of U.S. food, one-fourth of the country's hogs and one-fifth of its corn. Soybeans, cattle, and oats are important. Manufacturing mainstays are meat and other food-processing and agricultural equipment. Finance center Des Moines is second to Hartford, Connecticut, in insurance. Publishing and electrical products are increasing.

Visiting Iowa

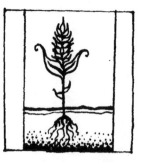

- May is Tulip Time in Pella and sees the Tulip Festival in Orange City, honoring Dutch immigrants and their beloved flower.

- Casino gambling is legal on riverboats on the Mississippi and Missouri rivers.

- Summertime exhibition games are played in Dyersville on the baseball field cut from a cornfield for the film *Field of Dreams*.

- Madison County boasts six famous, century-old covered bridges. In Winterset, actor John Wayne's birthplace is open for tours; Madison County Historical Complex includes 1850s' restorations.

- Davenport's Art Museum and the Putnam Museum of Science and Natural History offer excellent exhibits; nearby Buffalo Bill Cody Homestead and Buffalo Bill Museum cover the frontiersman/showman. Casino and sightseeing boats cruise the Mississippi.

- Kalona is center to the largest Amish community west of the Mississippi, with the Kalona Historical Village, Quilt and Textile Museum, and Mennonite Museum telling the story of these "plain people."

- Backbone State Park, northwest of Manchester, is named for the Devil's Backbone, the exposed rock that hikers climb; it also offers swimming, camping, cross-country skiing, and snowmobiling.

- Effigy Mounds National Monument protects 191 Indian burial mounds, many in bird and animal shapes. Trail hikes and guided tours are available.

- In Des Moines, visitors tour the Iowa State Capitol and Terrace Hill, the Victorian-era governor's mansion. Des Moines Art Center shows American and European paintings. There is a Center of Science and Industry; the city offers a zoo, botanical gardens, and an aquarium.

- Vesterheim Norwegian-American Museum in Decorah fills a city block with the story of these immigrants.

- Amana Colonies northwest of Iowa City are seven German villages with working crafters, museums, and restaurants. A nature trail follows the Iowa River. On 600 acres, nearby Living History Farms demonstrate farmwork of five different time periods.

KANSAS

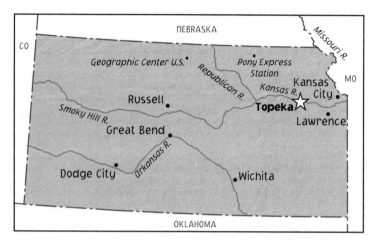

square miles: 81,815 (14th largest)
population: 2,688,418
density: 33 people per square mile
capital: Topeka
largest city: Wichita
statehood: January 29, 1861 (34th state)
nickname: The Sunflower State
motto: *Ad Astra per Aspera*
(To the Stars Through Difficulties)
bird: Western meadowlark
flower: native sunflower
tree: cottonwood

Land

Kansas rises gradually from prairie to high plains, 680' in the east to 4,039' at Mount Sunflower near the Colorado border. Millions of years ago an inland sea covered it, leaving some of the world's richest soil. Even though plains dominate, interesting geological formations include the mesalike Gypsum Hills, chalk spires of Castle Rock, and Horse Thief Canyon, a miniature Grand Canyon. Principal rivers are the Arkansas (flowing from Colorado) and the Kansas, which rises in western Kansas. North America's geographic center is in Osborne County, and the lower 48 states' geographic center lies near Smith Center.

Climate

Kansas's temperate climate experiences seasonal extremes, with January's average temperature 30°F (-1°C) and July's 78°F (26°C). Annual precipitation ranges from over 40" in the east to 20" in the west. The growing season runs from mid-April to mid-September.

People & History

Francisco Vásquez de Coronado and his Spanish explorers came from Mexico into central Kansas in 1541 to seek legendary cities of gold; but they found only a few Indian villages and plains filled with buffalo. As part of French-claimed land, the northeast of the Louisiana Territory received French fur traders in the 1700s. It came under U.S. control in 1803, its land explored but not settled by parties heading west.

In 1830 Kansas was part of the Indian Territory, set aside for tribes displaced elsewhere, until the Kansas-Nebraska Act of 1854 made two territories open to white settlers. Residents of Kansas determined whether their state would be free or slave, and immigrants from both South and North moved in to support their causes. During the 1850s, raids on abolitionist settlements by Southerners led to the term bleeding Kansas, but Kansas entered the Union as a free state. Immigrants from New England were joined by Germans, Russians, Bohemians, and Scandinavians.

Indian raids against settlers continued until the late 1870s, when most Indians were removed to Oklahoma. Eastern Kansas homesteaders built log cabins, but in the treeless west they cut prairie sod to stack for houses, known as soddies.

As railroads built westward across Kansas in the late 1860s and into the 1870s, one town after another experienced becoming a wild and woolly cowtown. Texas herds were driven to the current end of the railroad, and cowboys flush with pay supported its saloons. But that era lasted only a decade.

Mennonite immigrants in 1874 brought a good variety of wheat that flourished in Kansas's rich soil. The prairie was plowed and planted, and World War I's demand for wheat further enriched the state. The 1930s Dust Bowl, when winds stripped the soil, was a major setback. World War II saw a rise in aircraft manufacturing, mostly in Wichita, although agriculture continued in importance.

Industry

Kansas is the number-one wheat-producing state and a leader in flour milling and grain storage. Food processing, especially meat packing, is important. Kansas leads the nation in producing civilian aircraft. Thousands of oil and natural gas wells are scattered over the land.

Visiting Kansas

- At the University of Kansas in Lawrence, the Spencer Museum of Art exhibits Renaissance and Baroque paintings, 19th- and 20th-century American works, and graphic arts, including Japanese prints; the Natural History Museum's displays of Kansas birds, animals, and fossils include Comanche, the horse that survived the Battle of the Little Bighorn.

- Wichita's Old Cow Town Museum has restored or reproduced 44 buildings from the city's early days, 1865–80; Sedgwick County Zoo exhibits 2,000 animals in habitats from several continents; Wichita Center for the Arts includes a sculpture garden, exhibits, and performances; the Omnisphere Earth-Space Center offers hands-on exhibits about scientific principles.

- Mid-America Air Museum, at Liberal, holds 80 aircraft: military, civilian, and experimental, plus a wind tunnel and flight simulator. Elsewhere in town, Dorothy's House pays tribute to the fictional flying Kansan in *The Wizard of Oz*.

- Frontier Army Museum, Leavenworth, tells Civil War, Mexican War, and frontier Indian Wars history in exhibits and videos; visitors can leave from here on a self-guided tour of Fort Leavenworth, established in 1827.

- Topeka is Kansas's capital, with a French Renaissance limestone State House, Historic Ward-Meade Park, which is centered on an 1870s Victorian mansion furnished in that period and includes botanical gardens, a log cabin, a late 1800s schoolhouse, a livery stable, a drug store with working soda fountain, a general store, and a Santa Fe Railroad depot.

- Newton's Kauffman Museum gives the story of 1870s Mennonite immigrants from Europe, and it includes a homestead log cabin, 1875 farmhouse, 1886 barn, plus prairie wildflowers, woods, and grasses.

- Kansas Cosmosphere and Space Center, in Hutchinson, exhibits spacecraft and memorabilia, features an IMAX theater and laser show.

KENTUCKY

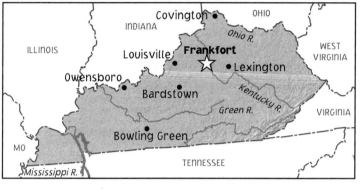

square miles: 39,728 (37th largest)
population: 4,041,769
density: 102 people per square mile
capital: Frankfort
largest city: Louisville
statehood: June 1, 1792 (15th state)
nickname: Bluegrass State
motto: United We Stand, Divided We Fall
bird: Kentucky cardinal
flower: goldenrod
tree: Kentucky coffeetree

Land

Kentucky's eastern quarter holds the Cumberland Mountains and Pine Mountain Range, home to eastern coal mines, with the state's highest point atop Black Mountain, at 4,139'. The Knobs to the West, a narrow horseshoe-shaped area of rounded hills and poor shale soil, touch at both ends of the Ohio River and enclose the Bluegrass country, named for the rich grass that supports world-famous horse farms. At the state's western edge is the Purchase, bordered by the Tennessee, Mississippi, and Ohio rivers—fertile hardwood forestland bought from the

Chickasaw Indians in 1818. East of it lies the Western Coal Field, where mining and farming mix. Filling in the rest of the state, the Pennyroyal region of rocky forested hills is home to small farms and many limestone caves.

Climate

Kentucky's January average temperature is 34°F (1°C), and July averages 77°F (25°C). Precipitation, distributed evenly around the calendar, is 48" in the south, 40" in the northeast.

People & History

Before whites arrived, Kentucky was a hunting and battle ground for Indians who lived to the north and south. The Cumberland Mountains barred exploration by east-coast colonists until a group led by Daniel Boone blazed a trail through the Cumberland Gap to the Bluegrass in 1775. During the Revolutionary War, British soldiers encouraged Indians to attack settlers, fulfilling Cherokee chief Dragging-Canoe's prediction that here would be a "dark and bloody land."

Following the War of 1812, settlers, mainly English and Scotch-Irish, flooded in from eastern states. French settlers moved up the Mississippi River—Kentucky's western border—most settling in the Louisville area. In the mid-1800s, German immigrants arrived via the Ohio River.

A few plantations in the Bluegrass and Pennyroyal used African American slave labor, but the legislature banned importation of slaves for resale in 1833. Many Kentuckians participated in the Underground Railroad, which helped escaped slaves move north and cross the Ohio River to freedom.

Kentucky did not secede from the Union during the Civil War, and more than twice as many men (90,000) fought for the Union as for the Confederacy (40,000). Only one major Civil War battle occurred here—at Perryville, where Confederate troops were defeated—but guerrilla warfare was widespread. At war's end, black citizens were given the vote, but Ku Klux Klan activity helped support their second-class citizenship.

After the war, many types of industry developed here. Dangerous conditions, then mechanization, led coal miners to seek labor union membership. Labor wars in the 1920s and 1930s resulted in violence and death and produced protest ballads that became part of American folk culture.

Industry

During the first half of the 1800s, railroads into eastern Kentucky led to development of coal mining, tobacco, and hemp (for nautical rope) production. Today Kentucky leads the nation in Thoroughbred horses, and seed of the horse country's famous bluegrass is also an important product. The state ranks second to West Virginia in coal from underground and strip mines. Whiskey was first made in Louisville in 1783, and Kentucky bourbon continues as a major source of income. Lumbering and furniture-making are important, and every baseball fan knows of the Louisville Slugger bat, still manufactured in Kentucky's largest city.

Visiting Kentucky

- Kentucky Derby Museum in Louisville is the next best thing to attending the annual May horse race: Here are Thoroughbred history, an audiovisual re-creation of Derby Day events, and interactive computer exhibits.

- Bardstown combines a walking tour of late-1700s buildings, Oscar Getz Museum of Whiskey History, and My Old Kentucky Home State Park honoring songwriter Stephen Foster—here a summertime musical about Foster's life and many works is often performed.

- Pop culture buffs will enjoy the National Corvette Museum at Bowling Green and Schmidt's Museum of Coca-Cola Memorabilia at Elizabethtown.

- The intentionally plain homes and furnishings of the Shaker religious sect can be seen at the restored Shaker Village of Pleasant Hill.

- White Hall Historic Site at Richmond preserves the estate of abolitionist Cassius Marcellus Clay.

- Abraham Lincoln Birthplace National Historic Site south of Hodgenville features a reconstruction of the 16th President's log cabin inside a stone memorial.

- Mary Todd Lincoln House in Louisville is the Georgian childhood home of the former first lady.

- Cumberland Gap National Historical Park marks the natural mountain break where Indians and white pioneers passed through the Allegheny Mountains.

- Mammoth Cave National Park, northeast of Bowling Green, includes 300 miles of known underground passages on five levels. Aboveground are self-guided nature trails.

LOUISIANA

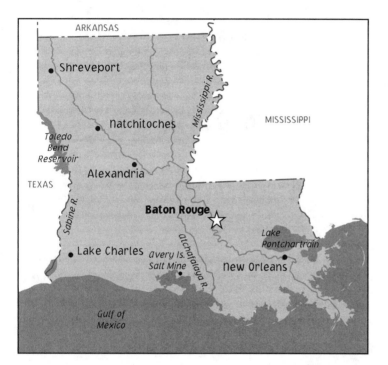

square miles: 43,562 (31st largest)
population: 4,468,976
density: 103 people per square mile
capital: Baton Rouge
largest city: New Orleans
statehood: April 30, 1812 (18th state)
nickname: The Pelican State
motto: Union, Justice, Confidence
bird: Eastern brown pelican
flower: magnolia
tree: bald cypress

Land

Louisiana's highest point, Driskill Mountain in the northwest, is 535' above sea level. New Orleans is protected by man-made banks (levees), and sits 8' below sea level. The Mississippi River reaches the Gulf of Mexico in an extensive delta. Its ancient floods—and those of the Red River to its west—created rich soils now protected by erosion-preventing tree farms. Human construction has harnessed the Mississippi into an unmoving channel, which causes erosion of the marshy, coastal beaches on the 397 miles of Gulf Coast. Wetlands of southern Louisiana are called bayous.

Climate

Louisiana's subtropical climate means mild winters and hot, humid summers with frequent showers. Annual precipitation averages 57". Hurricane season lasts from June through November. The average January temperature is 50°F (10°C), with July's average rising to 82°F (28°C).

People & History

Spanish conquistador Hernando de Soto led the first whites into future Louisiana, where he died of a fever in 1542. An estimated 15,000 Indians lived here then, most hunting and gathering, some farming in river bottoms.

The Spanish didn't create colonies—that was left to the French, beginning in 1699, seventeen years after René Cavalier Robert, Sieur de la Salle, came down the Mississippi and claimed the land around the river for France. They founded New Orleans in 1718; thirteen years later, Orleans Territory (roughly today's state shape) became a French crown colony. Control passed from France to Spain and back to France by treaties between 1762 and 1800. France gave the Baton Rouge area to Britain in the 1763 treaty that ended the French and Indian War; Spain conquered it during the American Revolution but later turned it over to France.

France sold Orleans Territory and Louisiana Territory (all land drained by the Missouri and Mississippi rivers) to the United States in 1803. Orleans Territory and the eastern Florida parishes (counties) united as the Louisiana Territory of the U.S. that same year. Louisiana became a state nine years later.

Settlement continued while colonial governments changed. Descendants of French colonists and Spanish settlers created a subculture called Creole. Thousands of Germans settled on the Mississippi north of New Orleans, an area called the German Coast. Acadians (French colonists the British expelled from Nova Scotia in 1755) also made their way to southern Louisiana. Their descendants are today's Cajuns. African Americans were brought as slaves to the cotton and rice plantations of inland Louisiana. Other ethnic groups included Dalmations (from today's Croatia), Italians, Hungarians, and more recently Cubans—giving Louisiana the richest cultural mix of the Deep South.

Although most citizens—small farmers who owned no slaves—didn't support seceding from the Union before the Civil War, political power belonged to owners of slave-worked plantations, who saw that Louisiana joined the Confederacy. After the war, former slaves, about one-third of the population, couldn't vote until the 1960s civil rights movement. But their cultural contributions were major—for example, New Orleans is now known as the birthplace of jazz music.

Despite a lumber boom in the early 20th century, Louisiana remained a poor, largely agricultural and raw-material-producing state until industrialization began with World War II.

Industry

Extensive oil deposits have been developed all around Louisiana, including offshore in the Gulf; the state is second only to Texas in U.S. oil production. The Gulf also makes Louisiana number one in volume of commercially caught seafood and fish. Natural gas, salt, and sulfur are among its chemical products. Crops include sugarcane, rice, soybeans, and some cotton; beef cattle are raised. Paper products come from the pine forests; sugar, rice, and seafood are processed. Peppers grown on Avery Island are made into Tabasco sauce. Shipping on the Mississippi River has always brought income, especially for Baton Rouge and New Orleans.

Visiting Louisiana

- Rip Van Winkle Gardens at New Iberia, built by 19th-century actor Joseph Jefferson (famous for playing Van Winkle), includes 25 acres of subtropical flora and a furnished Victorian home.

- The Louisiana Capitol in Baton Rouge is 34 stories tall, with an observation platform on the 27th floor. It was built during Huey Long's governorship (1928–32); he was assassinated within the building in 1935 and buried on its grounds.

- Jungle Gardens at Avery Island offers 200 acres of subtropical plants and trees and Bird City, a sanctuary where 20,000 visiting herons and egrets nest.

- Mardi Gras in New Orleans sees dozens of parades by krewes, organizations that prepare floats all year for the wild street festival leading to the beginning of Lent.

- A trip to New Orleans isn't complete without enjoying jazz, and Preservation Hall is where the most-traditional bands play.

- New Year's Day brings the Sugar Bowl to New Orleans's Superdome, otherwise home of pro football's Saints.

- Nottoway Plantation at White Castle reveals life before the Civil War. The mansion has 64 rooms furnished in the style of 1859, when it was built.

- Louisiana calls itself "Sportsman's Paradise" because of fishing, hunting, and boating opportunities throughout the state.

MAINE

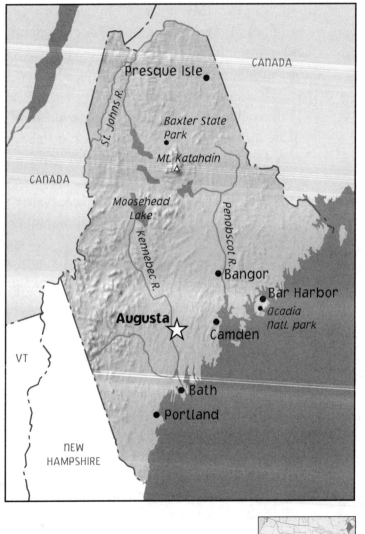

square miles: 30,862 (39th largest)
population: 1,274,923
density: 41 people per square mile
capital: Augusta
largest city: Portland
statehood: March 15, 1820 (23rd state)
nickname: The Pine Tree State
motto: Dirigo (I Direct)
bird: chickadee
flower: white pine cone and tassel
tree: white pine

Land

The largest New England state has four geographical regions. Coastal lowland reaches 20–30 miles inland from Eastport to Kittery along the 228 miles of mostly rocky Atlantic Ocean coast. Inland come the low interior hills, which also extend to the coast around Bar Harbor and Camden. The Longfellow Mountains, part of the Appalachians, include the state's highest point, Mount Katahdin, 5,268'. Five thousand rivers and streams are dominated by south-flowing Androscoggin, Kennebec, Penobscot, and Saint Croix, and by north-flowing Saint John and its tributaries

Allagash and Arroostock. On Mount Desert Island, 1,530' Cadillac Mountain is the highest point on the U.S. Atlantic coast.

Climate

Major daily and annual temperature changes mark Maine's climate, which differs between the north and the coast. January average temperatures are 10°F(-12°C) in the north and 20°F (-7°C) on the coast. July sees more even temperatures, averaging 68°F (20°C) statewide. Annual precipitation is 46" on the coast, 38" in the north.

People & History

Vikings may have visited the Maine coast about 1000 A.D., England's John Cabot claimed to have landed here in the 1490s, and expeditions from all European sea powers explored the coast and bays during the 1500s. Some nations created summer fishing stations on islands and opened fur trading in the 1600s. Iroquois raids, European diseases, loss of their land, and battles where they sided with the French killed much of the native Abnaki Confederacy Indian tribes (Passamaquoddy, Penobscot, and others). However, survivors' descendants live in the state today.

Samuel de Champlain and the Sieur de Monts founded today's Annapolis Royal in 1604, the first European settlement. France and England were the finalists in claiming the area, with England the winner in 1763 when it defeated France in eastern Canada.

Massachusetts Bay Colony took over Maine from its royal charter holder in 1652, and Maine was part of Massachusetts until its own statehood in the Missouri Compromise (see Introduction). Settlers came from England, Scotland, and Ireland and were joined in 1763 by large numbers of French-speaking Acadians driven from Nova Scotia by the British. In the St. John valley, bordering Canada, French is still spoken and used on signs. Germans and Italians also arrived, especially during the 19th century.

Maine boomed from the 1830s to the 1860s in fishing, shipbuilding, international ocean trade, timbering, ice harvesting, granite mining, and lime production. Lumber barons and sea captains built themselves grand wood-frame mansions along the coast. Portland's major deepwater port was downeast to ships leaving Boston, sailing downwind as they went east.

Industrialization was not as great here in the late 19th century as in southern New England; despite some factories, Maine was left with fishing and timbering and agriculture in those years. As many people moved elsewhere for work, its population grew slowly from 1860 to 1970. With few large valleys and mostly rocky land, Maine became a place of small farms and small, isolated towns in the days before modern highways.

Industry

Textile and shoe manufacturing have declined since the 1950s, but manufacturing continues as the main economic activity. With 90% of Maine's land covered by forest, the state is a leading paper producer, and lumber, fuelwood, furniture, and hardwood products are significant. Naval shipbuilding is less important than it once was, but still a major activity. Tourism is second in economic contribution from coastal and inland resorts as well as hunting, especially for deer and moose. Lobsters are the biggest portion of fishing income. Maine is a primary potato-raising state. From a single deposit in the southwest, the state is one of the U.S.'s main garnet producers.

Visiting Maine

• Bar Harbor became a resort town in the 19th century and is a lively one today. It offers access to an auto ferry to Nova Scotia and to Acadia National Park. The park, mostly on Mount Desert Island, covers nearly 42,000 acres, much of them donated by John D. Rockefeller Jr., and features a 20-mile ocean-view drive.

• From Fort Western, Augusta's first permanent building and now restored as a museum, American troops under Benedict Arnold marched on Quebec in the winter of 1775, but they failed to take the city from Britain. The

Maine Capitol, built 1828–31, was designed by Charles Bulfinch, who developed America's Federal style of architecture.

• York's Old Gaol Museum is in Maine's oldest public building, which was its jail from 1653 to 1860.

• Portland offers art museums, a children's museum, and the boyhood home of popular 19th-century poet Henry Wadsworth Longfellow. Nearby Portland Head Light, a lighthouse commissioned by George Washington and built in 1791, rises 101' above the ocean. Tate House, the city's oldest home, dates from 1755.

• Exhibits at Shore Village Museum, Rockland, include scrimshaw (bone and shell carvings made by sailors), lighthouse and life-saving artifacts, and model ships.

• Searsport's Penobscot Marine Museum exhibits ocean-fishing and whaling gear, ships' logs and sailing charts, and navigation instruments.

• Baxter State Park, north of Millinocket, surrounds Mount Katahdin.

MARYLAND

square miles: 9,774 (42nd largest)
population: 5,296,486
density: 542 people per square mile
capital: Annapolis
largest city: Baltimore
statehood: April 28, 1788 (7th state)
nickname: The Old Line State
motto: Fatti Maschii, Parole Femine (Manly Deeds, Womanly Words)
bird: Baltimore oriole
flower: black-eyed Susan
tree: white oak

Land

Maryland has 31 miles of coastline, mainly on Chesapeake Bay and the mouth of the Potomac River. East of Chesapeake bay is called the Eastern Shore, and west of the bay is called Maryland Main. The coastal plain has sandy soil in the south, but rich soil in the north. In the Piedmont Plateau, farther inland, are fertile soil and belts of clay used in brick-making. The Appalachian Mountains cross western Maryland, where Backbone Mountain is its highest point at 3,360'. Forming the state's southern border is the Potomac River.

Climate

Eastern Maryland has humid subtropical weather, whereas western Maryland's climate is continental. Average January temperature is 29°F (-2°C) west, 39°F (4°C) east; July averages are 68°F (20°C) west, 75°F (24°C) east. Precipitation is about 43" a year, falling evenly around the state.

People & History

Humans came into this land around 8000 B.C. as the glacier melted northward. By the time Europeans arrived, Algonquin nations lived in future Maryland. Threatened by Iroquois neighbors to the north, they welcomed white settlers as allies. But over six decades, whites drove them out.

In 1634, Englishman Leonard Calvert, Lord Baltimore's younger brother and holder of a royal land grant, led the first colonists to today's Blakistone Island in the Potomac. They created tobacco farms worked by indentured English servants (as of 1639, African slaves were added) and traded with the Indians.

The Calvert family promoted freedom of religion among Christian sects, which became Maryland law in 1649. That year, the port city Annapolis became the capital. Settlement expanded north and west, with Baltimore founded in 1729.

By the American Revolution in 1776, Calvert's was the only royal grant still in effect among the colonies, but most residents sided with rebellion. At war's end, the Continental Congress met at Annapolis, making it the U.S. capital for seven months beginning in late 1783.

After burning Washington, D.C., in the War of 1812, British troops headed for Fort McHenry in Baltimore harbor but were unable to capture it. Their all-day and all-night bombardment inspired an eyewitness, Francis Scott Key, to write "The Star-Spangled Banner," adopted as the U.S. national anthem in 1931.

Maryland was in the forefront of progress before the Civil War, building roads, canals, and the first passenger railroad, the Baltimore & Ohio. Although slavery was legal in Maryland, the state also prevented freed African Americans from being recaptured; more freed blacks lived here than in any other state just before the Civil War.

Residents were divided on the issue of Civil War, and Union troops occupied Baltimore and Annapolis. The war's bloodiest one-day battle occurred at Sharpsburg on Antietam Creek in September 1862, where nearly 8,000 men were killed or later died of wounds.

Maryland prospered in trade after the war, receiving raw materials through its ports and transporting them inland. The original English population had altered when Germans, including Jews, immigrated here in the late 1840s; now Italians, Greeks, Russian Jews, Poles, and Czechs settled throughout the state. Maryland soon had greater ethnic diversity than surrounding states. Southerners (including more African Americans) seeking economic opportunity also moved here after the Civil War. Many military bases built here helped support the economy from the 19th into the 20th centuries.

Industry

Maryland benefits from its nearness to Washington, D.C., with many federal government offices located here along with military sites that include Andrews Air Force Base, Goddard Space Flight Center, and Aberdeen Proving Ground; computer software and biotech research laboratories that work on federal contracts. Although ship and auto making have declined, they are still important along with steel manufacturing. Passenger trains create business along the corridor from Washington through Baltimore to Philadelphia. Its ocean port makes Baltimore one of the nation's main containerized freight and automobile arrival points. Following pollution-fighting efforts in recent years, Chesapeake Bay's commercial fishery improved, with crabs, clams, and oysters the main catches.

Visiting Maryland

- Chesapeake and Ohio Canal National Historic Park, reaching from Cumberland to Washington, shows the story of canals.

- The Preakness Stakes horse race in May is the centerpiece of Baltimore's annual Preakness Celebration; Babe Ruth House is the restored birthplace of the baseball legend; Poe House is where horror author Edgar Allan Poe once lived; Fort McHenry, built in the 1790s, is a National Monument. The National Aquarium features fish, reptile, and bird exhibits; the Maryland Science Center includes a museum, planetarium, and an IMAX theater. Opera, symphony, theater, art museums, and professional baseball and football are other Baltimore offerings.

- Among the many colonial buildings in Annapolis, the State House is the oldest U.S. capitol still used by a legislature; built in 1772, it was the national capitol from November 1783 to June 1784.

- St. Michaels' Chesapeake Bay Maritime Museum has ships afloat, a lighthouse, saltwater aquarium, and waterfowl exhibit.

- Assateague State and National Seashore is home to small wild horses known to have descended from animals that survived from a wrecked Spanish ship.

MASSACHUSETTS

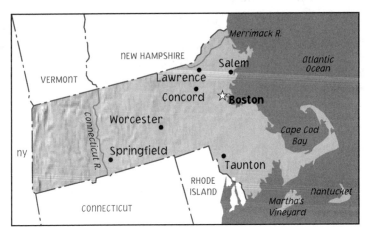

square miles: 7,840 (45th largest)
population: 6,349,097
density: 810 people per square mile
capital: Boston
largest city: Boston
statehood: February 6, 1788 (6th state)
nickname: The Bay State
motto: *Ense Petit Placidam Sub Libertate Quietem*
(By the Sword We Seek Peace, but Peace Only under Liberty)
bird: chickadee
flower: mayflower
tree: American elm

Land

Massachusetts has 300 miles of Atlantic coast and includes Cape Cod curling 65 miles into the ocean plus several islands—the largest two being Martha's Vineyard and Nantucket. Eastern Massachusetts' glacier-scoured land is hard and infertile, but the Connecticut Valley in the center is fertile, rolling plains. The land rises to the Berkshire Hills, where Mount Greylock, at 3,491', is Massachusetts' highest point. West of there lie the Berkshire valley and then the Taconic Mountains. The state's major rivers are the Connecticut, Charles, and Merrimack.

Climate

The coast is generally warmer and more moist than western Massachusetts, although the latter has severe winter snowstorms. Precipitation averages 42" a year statewide. January mean temperatures are 24°F (-4°C) west and 31°F (-1°C) coast, but July's means are closer between regions, 67°F (19°C) west and 71°F (22°C) coast.

People & History

The Massachusetts coast had been visited by Europeans (as early as 1003 by Vikings on Cape Cod), fished (during the 1500s), and mapped (by Samuel de Champlain in 1605 and John Smith in 1614), before colonists arrived in 1620. These Pilgrims spent their first winter uncomfortably aboard their ship, the *Mayflower*, which left in the spring. Pemaquid and Wampanoag Indians, who had lost thousands to an illness that also killed some Pilgrims, welcomed them. Colonists and Indians farmed together and greeted autumn's harvest with a thanksgiving celebration. Peace continued until 1675, when Prince Philip, the Wampanoag's chief, led Prince Philip's War, killing colonists and burning homes until his own defeat and death the following year.

The comparatively severe New England climate and the realities of plain frontier living removed the colonists from English influence. New ideas of freedom and equality shaped their attitudes. Massachusetts produced many leaders of the American Revolution, and its citizens supported the movement toward independence. In 1773, protesting yet one more British tax, Bostonians dumped the precious import—tea—into their harbor. Britain closed the port, but other colonies copied the Boston Tea Party.

Two years later, British troops, assigned to capture gunpowder from the provincial congress, killed minutemen (American volunteer soldiers) at Lexington and Concord, and the Revolutionary War began. In June 1775, when British troops took Breed's Hill above Boston (the battle remembered as Bunker Hill, a nearby site), they lost more than 40% of their force. The British left Boston in 1776.

Massachusetts was a leader in the Industrial Revolution as well, its factories dating from the 1640s. Lowell, settled in 1653, became a textile center in the early 1800s, its mills powered by the Merrimack River. Beverly was a shoemaking center. The state was an arsenal to the nation during all its wars including the Revolution. By the early 20th century, many factories had become outdated, and the state largely lost its shoe and textile industries.

As factories increased, Irish immigrants fleeing famine in the 1840s began to arrive and work in them. Large numbers of Italians came in the early 20th century. Many Portuguese moved to Fall River and New Bedford. Immigrants from around the world today make Massachusetts more ethnically diverse than any other Atlantic state but New York.

Industry

Manufacturing is no longer the main employer in Massachusetts, but products include computers, other electronic equipment, surgical instruments, and printing. Service industries are the main employer and include banking, trade (Boston is New England's largest retail center), and real estate. Tourism is very important to the Boston area, Cape Cod and the islands, and the Berkshire Hills. Research facilities and many outstanding private universities also contribute to the economy.

Visiting Massachusetts

• Boston's Freedom Trail guides visitors to 16 historic sites along three miles; the Black Heritage Trail on Beacon Hill shows nine sites including an Underground Railroad "station" that aided escaping slaves; at Charlestown Navy Yard, War of 1812 ship USS *Constitution*—known as Old Ironsides—is docked; the Boston Pops Orchestra performs free outdoor summer concerts on the Esplanade (a park on the Charles River) and the Boston Symphony has a winter concert season; professional teams play baseball,

basketball, and football; the Museum of Fine Arts exhibits objects from around the world and over the last 5,000 years; the Museum of Science's many exhibits and facilities include the Hayden Planetarium.

• Plimouth Plantation, at Plymouth, shows a Pilgrim village as it would have been in 1627; it includes the reconstructed ship *Mayflower II*, which sailed from Plymouth, England, in 1957.

• Between Lexington and Concord, Minuteman National Historical Park honors volunteer American Revolution soldiers who drilled to be ready to fight on a minute's notice.

• Salem makes much of its witchcraft trials of 1692, with a Witch Museum and Witch House. It also has the childhood home of author Nathaniel Hawthorne.

• Cape Cod, Martha's Vineyard, and Nantucket Island offer summer resorts and sandy beaches and moors.

MICHIGAN

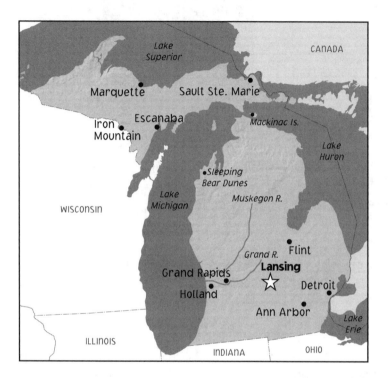

square miles: 56,804 (23rd largest)
population: 9,938,444
density: 175 people per square mile
capital: Lansing
largest city: Detroit
statehood: January 26, 1837 (26th state)
nickname: The Wolverine State
motto: *Si Quaeris Peninsulam Amoeam Circumspice*
(If You Seek a Pleasant Peninsula, Look Around You)
bird: robin
flower: apple blossom
tree: white pine

Land

Michigan occupies two peninsulas (an upper and a lower one) that touch four of the five Great Lakes, giving it 3,288 miles of coastline, including islands. Mount Arvon on the Upper Peninsula is the state's highest point at 1,979', but the peninsula's eastern lowlands average only 700'. The Lower Peninsula is mostly level but rises to 1,700' on glacial moraine near Cadillac. Lowest elevation is 572' on Lake Erie.

Climate

Winds from the Great Lakes cool the summers and warm the winters, making Michigan's climate more moderate than those of other north-central states. January's temperatures average 20°F (-7°C), July's average 69°F (21°C). Annual precipitation of 32" includes winter snowfall.

People & History

Ottawa Indians who were already here when French explorer Étienne Brulé arrived in 1622 had developed canoes in which they traveled the Great Lakes. Huron Indians of the southeast were farmers, and Miami and Potawatomi people lived in the vast forests. Chippewas and the Ottawas were important suppliers for French fur traders, who founded their first settlement, Sault Ste. Marie, in 1668. Thirty-three years later, Antoine de la Mothe Cadillac founded Detroit, France's most important post on the Great Lakes.

England won Michigan from France in 1760, and in 1763 made it part of Canada. During the Revolutionary War, British troops based in Detroit raided the western frontier until Americans captured the town in 1779. Indian opposition to settlement continued until 1794 and their defeat in the Battle of Fallen Timbers in Ohio. After trying to keep major posts, including Detroit, Britain finally left Michigan in 1796, and American settlement was quick.

Still, Michigan Territory's first governor surrendered Detroit to Britain during the War of 1812. The following year's Battle of Lake Erie won it back for the U.S.

Farmers, including Germans, Dutch, and escaped or freed African American slaves came into Michigan in the 1830s through the 1850s. Finns settled on the Upper Peninsula, where iron and copper were found in the 1840s. In the 1890s, many Poles came to live in Michigan's cities.

Detroit, Lansing, and Flint became the centers of automobile manufacturing in the first years of the 20th century. Henry Ford's idea of mass producing rather than custom building cars led to lower prices that allowed more people to buy them. The 1930s Depression was very hard on the state because a car still was not a necessity. During World War II, Detroit turned to war production, calling itself the Arsenal of Democracy. International changes in the industry have altered, but not destroyed, Michigan's major industry.

Industry

Twenty-two percent of American automobiles are manufactured in Michigan, their factories employing one-third of its workers. Midland is a chemical-making city. Cereals and food products are important, especially in Battle Creek. Grand Rapids and Muskegon are furniture-making centers. Although not a major agriculture state,

Michigan leads the nation in raising tart cherries, blueberries, and cucumbers. Metal ores are mined in the Upper Peninsula, including iron and some gold.

Visiting Michigan

- Isle Royale National Park's 400 islands combine wilderness and water for camping, hiking, fishing, and nonmotorized boating.

- At Sault Ste. Marie, visitors can view the Soo Locks of St. Mary's Falls Ship Canal from a 210' observation tower or in a boat passing through the locks. A maritime museum now fills the Valley Camp, a Great Lakes freight ship.

- Dearborn's Henry Ford Museum is about America, and Greenfield Village recreates life, crafts, and inventions from the turn of the 20th century.

- Automobiles are banned on Mackinac Island, which visitors reach by ferry. Walking, bicycling, horseback, and horse-drawn carriages are the ways to explore this state park that recalls the fur-trade era.

- Detroit, automobile-making center of the U.S., includes museums ranging from Detroit Institute of Arts to the Motown Museum, Detroit Historical Museum to Museum of African American History. A natural-habitat zoo, conservatory, freshwater aquarium, and Dossin Great Lakes Museum are on Belle Isle in the Detroit River. The Detroit Zoo in Royal Oak is a world-class facility. Professional teams play baseball, football, hockey, and basketball.

- Lansing, the state's capital, is a Victorian masterpiece. It is home to the Michigan Historical Museum, Women's Historical Center and Hall of Fame, and R. E. Olds Transportation Museum.

- In the aptly named city of Holland, tulip time is celebrated in May with tours of its tulip bulb fields and wooden shoe factories year-round. The only authentic working Dutch windmill in the U.S. is in its city park.

MINNESOTA

square miles: 79,610 (12th largest)
population: 4,919,479
density: 62 people per square mile
capital: St. Paul
largest city: Minneapolis
statehood: May 11, 1858 (32nd state)
nickname: North Star State
motto: *L'Étoile du Nord* (North Star)
bird: common loon
flower: showy lady's slipper
tree: Norway pine

Land

Near North America's center, Minnesota holds the Mississippi River's headwaters, borders the Great Lakes (which drain into the Atlantic Ocean), and sees the Red River of the North and Rainy River flow north to Hudson Bay. Part of its Canada border reaches farther north than any other land in the lower 48. Glaciers gouged Minnesota, leaving thousands of lakes, nearly 11,000 of which are larger than 25 acres. The state's highest point, Eagle Mountain, reaches 2,310', only 1,700' feet above Minnesota's lowest point along the 160 miles of Lake Superior shore.

Climate

In the northern part of the temperate zone, Minnesota has regional climates from moist Great Lakes on the northeast to semiarid Great Plains on the west. Northern Minnesota's January average daily high is 5°F (-5°C), in the south, 25°F (-4°C). July average daily highs are 80°F (27°) northwest and 85°F (29°C) south. Average annual precipitation is more than 30" in the southeast, but less than 20" in the northwest.

People & History

Chippewa Indians in the north and east and Sioux in the south and west lived here when the first French explorers arrived in the mid-17th century. Later, the Chippewa pushed the Sioux southwest onto the prairie and occupied the northern forests.

French voyageurs (fur trappers) created the first white settlement at their portage (overland route) around the falls on the Pigeon River. In 1819, the military presence at Fort Snelling (located in today's St. Paul) led to the first U.S. settlement. Most immigrants in the next few decades came to log the vast forests. Lumber mills were built at Stillwater and at St. Anthony (which merged with Minneapolis in 1872).

The greatest rush of settlers came in the 1880s, when advertising in Scandinavian countries brought farmers and their families to homestead the West and Southwest. Wheat production soon led the state to be a flour-milling center.

Iron was discovered in the Vermillion Range in 1884, but much larger amounts found six years later in the Mesabi Range boosted the growth of nearby Great Lakes port cities of Superior and Duluth.

Farm mechanization in the 20th century meant that farms grew larger and employed fewer people. Dairy farming has increased in importance since the 1940s. As more people were employed in nonfarm jobs, about half the state's people lived in the Twin Cities (Minneapolis-St. Paul) area by the 1970s.

Industry

Service industries make up 70% of Minnesota's economy, including outdoor tourism both in summer and winter and medical services at Rochester's Mayo Clinic and the University of Minnesota in Minneapolis. The Mesabi Range is the U.S.'s main source of domestic iron ore. St. Cloud produces granite for monuments; the Minnesota Valley, granite for buildings. Minnesota is in the top 10 states for agriculture, producing grains, sugar beets, soybeans, cattle, hogs, and turkeys. Food processing includes a large milk, butter, and cheese industry. Manufactured goods include computers, industrial machinery, printing and publishing, wood products, and surgical instruments. Lake Superior ports are important transport centers.

Visiting Minnesota

• Bloomington is home to the Mall of America, the world's largest shopping center, with 400 stores, theaters, nightclubs, and an amusement park with roller coaster.

• Pipestone National Monument at Pipestone preserves a place sacred to Indians of the upper plains, who still quarry hard red stone here to carve peace pipes, which they have done for centuries.

• Minneapolis is an upper Midwest cultural hub, with the Guthrie Theater; Minneapolis Orchestra; Walker Art Center's modern art; Minneapolis Institute of the Arts exhibiting ancient to modern works; many small dance and theater companies; and the Planetarium at the Minneapolis Public Library. Parks with water-based recreation on lakes and the Mississippi River shore abound. Professional sports include baseball, basketball, and football.

• St. Paul, the other half of the Twin Cities, houses the St. Paul Chamber Orchestra in the Ordway Music Theatre; railroad magnate James J. Hill helped build the Cathedral of St. Paul; the Landmark Center opened in 1978 for performing arts and cultural events; historic Fort Snelling recreates the frontier post.

• Sauk Centre has the boyhood home of Nobel Prize novelist Sinclair Lewis, and the town was his model for the novel *Main Street*.

• Boundary Waters Canoe Area is the only national park that is preserved solely for canoeing.

• Voyageurs National Park offers summer and winter recreation, camping and lodging in forests surrounding fifty lakes.

• Rochester offers tours of the Mayo Clinic and of Mayowood, the estate of the clinic's founding brothers.

MISSISSIPPI

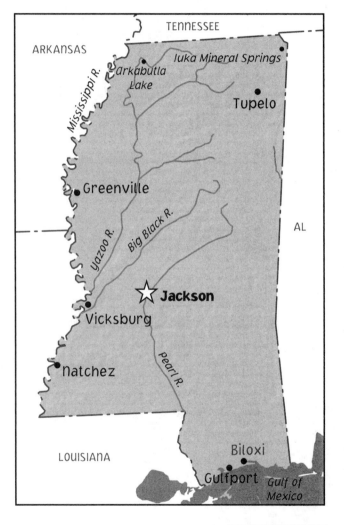

square miles: 46,907 (32nd largest)
population: 2,844,658
density: 61 people per square mile
capital: Jackson
largest city: Jackson
statehood: December 10, 1817 (20th state)
nickname: The Magnolia State
motto: *Virtute et Armis* (By Valor and Arms)
bird: mockingbird
flower: magnolia
tree: magnolia

Land

Mississippi takes its name from its western border, North America's greatest river, called by Algonquin-speaking Indians, "Father of Waters" (*misi* = big; *sipi* = water). The Delta region of northwestern Mississippi has rich black soil left by eons of flooding by the Mississippi and Yazoo rivers. The Tennessee Hills of the extreme northeast include Woodall Mountain, the state's highest point at 806'. South of these regions lie Central Prairie and Black Prairie, once home to cotton plantations. South Mississippi, below Jackson, is the Piney Woods region of virgin pine forests.

The Pearl River and Tennessee-Tombigbee Waterway are also important to shipping. Mississippi has 44 miles of coastline on the Gulf of Mexico.

Climate

Summer is hot and rainy, autumn crisp, and winter mild. July's average temperature in Mississippi is 80°F (27°C), with January's average 46°F (8°C). Snowfall is very rare, but annual precipitation averages 50". Hurricane season on the coast is from June to October.

People & History

Ancestors of historic Native Americans, the temple-mound–building Mississippian Indians lived here in ancient times. Emerald Mound off the Natchez Trace Parkway at Milepost 10.3 is one of the largest surviving temple mounds.

Hernando de Soto of Spain first saw the Mississippi River at Sunflower Landing near Clarksdale in May 1540. But it was France who claimed and began to colonize this land. Pierre Le Moyne, Sieur d'Iberville, founded the first permanent white settlement at today's Ocean Springs in 1699. He founded Natchez seventeen years later. But few whites settled here until after 1763, when France ceded land east of the Mississippi River to Britain after the French and Indian War. Most settlers came from Britain's American colonies, and Mississippi's white population is primarily of British and northern European descent. Mississippi Territory was created in 1798, with lands added in 1804 and 1813 filling out its present shape. Statehood came in 1817.

When the 13,000 Choctaw Indians were ordered to move to Indian Territory (Oklahoma) beginning in 1830, almost half of them (7,000) managed to stay in Mississippi. Today their reservation is west of Philadelphia. The last of the Southern Indian tribes, the Chickasaws, left for Indian Territory in 1837–38.

Mississippi was the second state to secede from the Union and join the Confederacy in 1861. Vicksburg became a major Union target in 1862 because capturing it would divide the South and open the Mississippi River to Union ships. The city used slaves to build fortifications and held out until General Ulysses Grant surrounded and laid siege to it in May 1863. General John C. Pemberton's 30,000 Confederate troops could not escape, nor could aid reach them, and he surrendered on July 4.

The war ruined Mississippi's economy and much of its land; conditions from war's end in 1865 through the eleven years of Reconstruction grew worse. Former slaves, the majority of the population, became sharecroppers in debt to their landlords. A new constitution in 1890 denied civil rights to ex-slaves. The civil rights movement of the 1960s met with much violence here before it ended racial segregation.

Industry

Mississippi's pine forests, once cut recklessly, now are carefully tended and support the paper- and wood-products industry. Clothing and textiles are a major industry, and cloth and wood combine to make Mississippi one of the U.S.'s main makers of upholstered furniture. Cotton, rice, soybeans, corn, and wheat are important crops, along with pecans. The Gulf supports extensive commercial fishing that includes shrimp and oysters.

Visiting Mississippi

- Florewood River Plantation State Park at Greenwood shows plantation life in the 1850s; in late summer, visitors can try their hands at picking cotton.

- Casino gambling can be found throughout the state.

- Natchez has preserved 500 antebellum buildings and welcomes visitors into many homes during "pilgrimages" in March, October, and at Christmas. There's also a self-guided walking tour to the sites.

- Natchez Under-the-Hill recalls another side of river life, but today, the former hangout of gamblers and thieves houses shops and restaurants.

- Natchez Trace Parkway is a National Park along a highway that developed from an Indian trail from Natchez to Nashville, Tennessee. Interpretive signs, wayside exhibits, and nature trails fill the Trace's 500 miles.

- Jackson, capital and largest city, is home to the Mississippi Museum of Art, the Jackson Zoological Park (with a petting zoo), the Old Capitol (used 1838-1903) that houses the Mississippi Historical Museum, and many festivals and events around the calendar.

- Vicksburg National Military Park, on 1,700 acres, tells the story of the Civil War siege and also holds many monuments.

- Beach Boulevard offers a drive along the Gulf Coast, and the resort town of Biloxi with its blend of French, Spanish, and Old South culture.

MISSOURI

square miles: 68,886 (19th largest)
population: 5,595,211
density: 81 people per square mile
capital: Jefferson City
largest city: Kansas City
statehood: August 10, 1821 (24th state)
nickname: The Show Me State
motto: *Salus Populi Suprema Lex Esto*
(The Welfare of the People Shall Be the Supreme Law)
bird: bluebird
flower: hawthorn
tree: dogwood

Land

Dividing the state's northern third from its southern two thirds, the Missouri River flows west to east across Missouri. North of it lie plains and prairie, and rolling hills. To the south, hills are steeper, with streams cutting deep, narrow valleys. The Ozark Mountains extend from St. Louis into Arkansas and Oklahoma. Their highest point in Missouri is Taum Sauk Mountain, 1,772'. Southeastern Missouri is in the flood plain of the Mississippi River, the state's eastern border.

Climate

Missouri's average January temperature is 30°F (1°C); July average is 78°F (26°C). Summer temperatures of 100°F (38°C) occur statewide, but northwest summers are generally cooler. Annual rainfall is 35" north-northwest to 45" southeast, one third falling in April through June. Tornadoes can occur, usually in warm months.

People & History

By the time Europeans arrived in the late 1600s, Siouxan Indians such as the large Osage tribe and smaller Missouri had moved here from the east. French lead miners and hunters came into northeast Missouri about 1735, followed by French fur traders. Spain gained control of vast Louisiana Territory (which included future Missouri) in 1763.

After the American Revolution, American settlers threatened the Spanish colony. In 1800, Spain gave Louisiana Territory back to France, which sold it to the U.S. three years later. Missouri became a base for America's westward expansion: The Lewis and Clark Expedition wintered near St. Louis in 1803–04; St. Joseph became the place to buy supplies for the Santa Fe, Oregon, and California trails, and the eastern terminus of the Pony Express.

Missouri's application for statehood led to the Missouri Compromise (see Introduction) prohibiting slavery in western states. The southeast "bootheel" had a cotton plantation economy, but elsewhere residents—from northeastern states, and people from Germany, Ireland, and England—did not own slaves. When the Civil War began, Missouri stayed in the Union and sent four times as many soldiers to its army as to the Confederacy's.

Missourians had cleared the land of trees—except in the Ozark Mountains—and cultivated small family farms. After the war until 1900, Kansas City and St. Louis grew with immigrants matching earlier groups, and later Italians, Greeks, Poles, and Jews. Industry developed in cities, flourishing after World War II. Agriculture remained important, although farm quantity decreased as farms were made larger.

Industry

Missouri is among the top ten states raising soybeans, corn, winter wheat, cotton, and hay. Most of the corn feeds beef cattle, and the dairy industry also is important. Manufacturing accounts for 20% of the economy, but Missouri ranks third nationally in automobile production. Spacecraft and railroad equipment are made here. Food processing includes brewing, flour milling, and meat packing. Tourism is significant in the Ozarks, where many small towns depend upon it. St. Louis and Kansas City are national railroad hubs, and St. Louis on the Mississippi and Missouri rivers is the nation's major inland port.

Visiting Missouri

- The scenic Katy Trail State Park for hiking (and in some sections, bicycling) follows the Missouri River for 200 miles.

- In Jefferson City, the Missouri Capitol dates from 1918 and includes paintings by Thomas Hart Benton and N. C. Wyeth. Jefferson City Landing State Historic Site preserves buildings important in mid-19th century river trade.

- Mark Twain Boyhood Home and Museum, Hannibal, is the restored 1843 home of Samuel Clemens. Exhibits include Norman Rockwell paintings for special editions of Twain novels.

- Branson offers nearly 40 music theaters, mostly country-western but including other musical styles. Nearby Shepherd of the Hills Homestead, based on the Harold Bell Wright novel, is a working homestead with gristmill and sawmill.

- St. Genevieve, settled by 1735, offers a self-guided walking tour among early 19th-century merchants' mansions and older buildings.

- Meramac Caverns, near Stanton, has five underground levels and a history that includes hideout for Jesse James' gang.

- St. Louis grew from a 1764 fur-trade post where the Missouri River flows into the Mississippi. Gateway Arch, 640' high, at Jefferson Expansion National Memorial, marks the post's site; the Museum of Westward Expansion on the grounds covers the Lewis and Clark Expedition and natural and cultural history.

The St. Louis Science Center includes hands-on exhibits. Missouri Botanical Gardens' 79 acres present English, Japanese, and other styles. The symphony, zoo, opera, theater, professional sports teams, riverboat casinos, and Six Flags St. Louis amusement park are among entertainment opportunities.

- Kansas City grew from a fur post at the state's west edge. Today the wheat-trading and industrial center offers the Toy and Miniature Museum; Thomas Hart Benton Home and Studio; the Nelson-Atkins Museum of Art's collections from ancient to contemporary, the Kansas City Museum about plains life, Osage Indians, and the city; the Negro Leagues Baseball Museum; a zoo, theater, professional sports, riverboat casinos.

- Independence was the start of the Santa Fe Trail, Oregon, and California trails and was home to Harry S. Truman. Museums cover all their stories.

MONTANA

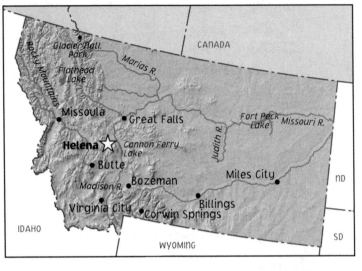

square miles: 145,552 (4th largest)
population: 902,195
density: 6 people per square mile
capital: Helena
largest city: Billings
statehood: November 8, 1889 (41st state)
nickname: The Treasure State
motto: *Oro y Plata* (Gold and Silver)
bird: Western meadowlark
flower: bitterroot
tree: ponderosa pine

Land

The Rocky Mountains fill Montana's western two-fifths, rolling to level Great Plains in its eastern three-fifths. The Missouri River's headwaters are in Montana; rivers east of the Continental Divide flow toward the Atlantic Ocean, and rivers west of it to the Pacific; some Glacier National Park rivers flow north into Hudson Bay feeders. Montana's lowest point is 1,800' where the Kootenai River flows out its northwest corner; and Granite Peak in the Beartooth Range, at 12,799', is its highest.

Climate

Semiarid Montana receives 15" average annual precipitation, although west of the Continental Divide is more moist Pacific Northwest climate. January's average temperature is 18°F (-8°C), and July's is 68°F (20°C).

People & History

Future Montana's plains were home to Blackfeet, Cheyenne, Sioux, and Crow Indians, and its mountains to Salish, Kootenai, and Pend d'Oreille. French fur traders arrived by the 1740s, but the first recorded whites were the Lewis and Clark Expedition in 1805. Catholic missionaries created settlements in the 1840s.

Rich discoveries of placer (free-floating) gold in 1862 and 1863 brought a mining rush to southwestern Montana. Farmers and ranchers soon followed. In 1866, the first cattle drive from Texas brought cattle into central Montana's grasslands, where large, free-ranging herds thrived until a severe winter (1886–87) killed 90% of the animals. After that, smaller (and fenced) ranches raised hay for winter feed.

After placer gold dwindled, hard-rock underground mines opened. In Butte, miners were disappointed to find more copper than gold, but by the 1880s newly developed electricity called for copper wiring. Butte, atop "the richest hill on earth," dug mines nearly a mile under-ground. European immigrants flocked to jobs.

Wheat-raising developed on the plains in the 1910s, when unusually wet years aided crops. U.S. citizens and immigrants were drawn to homestead Montana, but drought conditions and low food prices in the 1920s ruined many. Banks failed and people moved away; survivors bought smaller claims, creating vast farms.

The Great Depression harmed mining, continuing drought harmed farming, but World War II's demands brought recovery. Oil was found in eastern Montana in 1951 and again in 1967, and Billings became a refinery and financial center.

Industry

Service industries, including government, finance, real estate, and tourism, account for most of the economy. Montana leads the nation in mining talc and vermiculite (used in insulation) and it holds the U.S.'s only platinum mine. Underground gold mining continues; copper is mined from a huge open pit in Butte. Extreme western Montana's forests yield pulp timber. Two-thirds of the land is given to ranches that raise mostly beef cattle. Manufacturing is limited to processing raw materials.

Visiting Montana

• Glacier National Park offers backcountry hiking amid glacier-topped peaks that are home to grizzly bears and bighorn sheep. Going-to-the-Sun Highway's two lanes are carved into the mountainside. Guided bus and lake boat tours, as well as short loop trails are available.

- Little Bighorn Battlefield National Monument honors the fallen of both sides in the 1876 battle in which Sioux and Cheyenne annihilated part of the Seventh Cavalry led by Colonel George Custer. Guided bus tour and self-guided driving tour are ways to see the large site.

- Great Falls is home to Lewis and Clark National Historic Trail Interpretive Center that overlooks five Missouri River waterfalls where their famous expedition needed a month to pass this point in 1805. C. M. Russell Museum holds Western art, and Russell's log cabin studio just as the artist left it.

- Across from the capitol in Helena, the Montana Historical Society has Russell art, and it presents exhibits on state prehistory and history. Holter Museum of Art exhibits contemporary works. The main street's name, Last Chance Gulch, recalls that miners who struck gold here took one last chance before their supplies ran out.

- Virginia City is half small county-seat, half preserved gold-rush town from 1863 whose buildings are a state museum. Nearby Nevada City holds buildings moved from other sites on Montana's gold frontier.

- Museum of the Plains Indian, at Browning on the Blackfeet Reservation, covers Northern Plains buffalo-hunting life.

- Grant-Kohrs Ranch National Historic Site at Deer Lodge is the working cattle ranch of a frontier cattle baron.

- Bighorn Canyon National Recreation Area south of Billings has boating on a lake surrounded by canyon walls, fishing, and camping. Nearby Pryor Mountain Wild Horse Range protects mustangs; it allows hiking, scenic drives, biking, and camping.

- At National Bison Range in Moiese, visitors stay in their cars to drive among free-ranging buffalo, antelope, elk, deer, and coyotes.

NEBRASKA

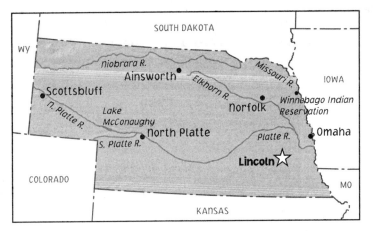

square miles: 76,872 (15th largest)
population: 1,711,263
density: 22 people per square mile
capital: Lincoln
largest city: Omaha
statehood: March 1, 1867 (37th state)
nickname: The Cornhusker State
motto: Equality Before the Law
bird: Western meadowlark
flower: goldenrod
tree: cottonwood

Land

Unforested Nebraska reaches from the Central Lowland into the Great Plains, from 840' in the southeast to 5,426' in Kimball Country in the northwest. The Missouri River forms its eastern border. Nebraska rises from the dissected till plains (river-cut, rich-soiled land) of its eastern fifth to the Great Plains on the west. In Central Nebraska, the Sand Hills are the largest area of sand dunes in North America, their grass covering provides cattle feed.

Climate

Nebraska experiences hot summers and cold winters, with precipitation that is greater in the east than in the west. The statewide average annual precipitation is 22". The January average temperature is 23°F (-5°C), and the July average 76°F (24°C).

People & History

Prehistoric people lived on this land as long ago as 8000 B.C. In historic times, eastern and central Nebraska held the homes of Pawnee, Omaha, Ponca, and Oto Indians, while the buffalo-hunting Sioux, Arapaho, Cheyenne, and Comanche roamed its west.

French fur traders along the Missouri reached Nebraska in about 1700, and Spaniards explored here twenty years later. France lost its claim to Spain in 1763, but bought the vast drainage of the Missouri River back from Spain in 1801. Two years later, the U.S. acquired those lands in the Louisiana Purchase. Traveling upstream on the Missouri, the Lewis and Clark Expedition in 1804 was the first to seriously explore Nebraska's eastern border. After his 1819–20 exploration, American Stephen Long said that Nebraska was part of the Great American Desert and not suitable for farming.

Future Nebraska later became part of Indiana Territory (1804–5), then part of Louisiana Territory until 1812, Missouri Territory to 1821, and Indian Territory (where white settlement was banned) until 1854. Then the Nebraska Territory was created, which included future Colorado, the Dakotas, Idaho, Montana, and Wyoming. From the 1840s, it was a place to pass through for emigrants on the Oregon, California, and Mormon wagon trails.

Beginning in the 1840s, towns first developed along the Missouri River to serve the fur trade and steamboat traffic headed for the northwestern frontier. When the Union Pacific Railroad, the first transcontinental rail line, was completed in 1869, inland development began. Now, farmers proved Stephen Long's opinion incorrect. Cattle ranching made Omaha an important meat-packing city, and other industries later developed there.

Depressions in the late 1860s, 1890s, and 1930s seriously affected agriculture-dependent Nebraska. In the 1890s, the Populist Party became important as it fought high transportation costs and low crop prices. The Great Depression of the 1930's—heightened by the Dust Bowl climate conditions that blew away topsoil—was very bad for Nebraska until federal price-supports for crops were created. World War II's food demands led to an economic recovery, and steady if not large growth since, despite a 1980s recession.

Industry

Nebraska is more of a farming state than any other, with 95% of its land used for agriculture. In the east, corn and soybeans are raised, north-central Nebraska is cattle-ranching country, and the west holds huge wheat farms. Omaha is a financial and service center for the Midwest; it is also a major meat-packing and grain market city.

Visiting Nebraska

- Lincoln is crowned by a 32' bronze statue, the Sower, atop the Nebraska Capitol; the grounds include a memorial to Abraham Lincoln by Daniel Chester French, designer of the Lincoln Memorial in Washington, D.C.

- Chimney Rock National Historic Site, east of Scottsbluff, holds the natural 325' clay monolith that told Oregon Trail travelers their trip was one-third over. Visitors can try out Conestoga wagon travel on three-day treks.

- Omaha offers Joslyn Art Museum's Western art collection that includes some of the earliest paintings of Plains Indians, works by Indian artists, and works by Frederic Remington. Western Heritage Museum, housed in a restored train depot, exhibits Union Pacific Railroad and Abraham Lincoln items. Henry Doorly Zoo and Lied Jungle include a rain-forest environment, open-air aviary, and tunnel leading viewers through the shark exhibit.

- Homestead National Monument of America, west of Beatrice, marks one of the first homestead claims filed. It includes a visitor center, original foundations and grounds, a one-room school, a homesteader cabin, plus a natural tallgrass prairie with walking trails.

- Agate Fossil Beds National Monument, north of Scottsbluff, holds fossils of early mammals from 19 million years ago, which can be viewed from a nature trail. Buffalo and native prairie grasses inhabit the park.

- Fort Kearny State Historical Park, Kearney, reconstructs the 1848 fort that housed troops protecting Oregon Trail wagon trains.

NEVADA

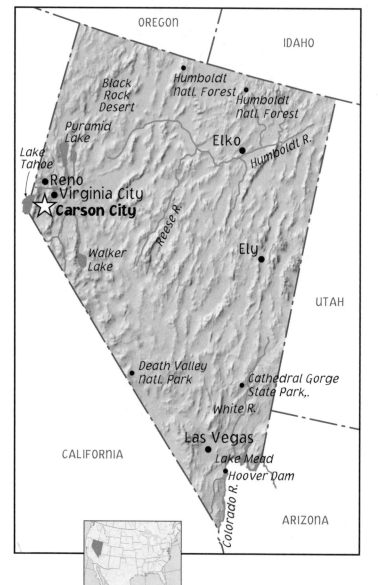

OREGON

IDAHO

Black Rock Desert

Humboldt Natl. Forest

Humboldt Natl. Forest

Pyramid Lake

Lake Tahoe

Elko

Humboldt R.

Reno

Virginia City

Reese R.

Carson City

Walker Lake

Ely

UTAH

Death Valley Natl. Park

Cathedral Gorge State Park,.

White R.

Las Vegas

CALIFORNIA

Lake Mead

Hoover Dam

Colorado R.

ARIZONA

square miles: 109,826 (7th largest)
population: 1,998,257
density: 18 people per square mile
capital: Carson City
largest city: Las Vegas
statehood: October 31, 1864 (36th state)
nickname: The Silver State
motto: All for Our Country
bird: Mountain bluebird
flower: sagebrush
tree: bristlecone pine and piñon

Land

Most of Nevada is in the Great Basin with a small part of Columbia Plateau in the northeast and the Sierra Nevada southwest of Carson City. Boundary Point southwest of Tonopah is the state's highest point at 13,143'; its lowest is 470' on the Colorado River in the southeast.

Climate

Nevada has long, hot summers, with a July average temperature of 73°F (23°C), and short, mild winters with January temperatures averaging 30°F (-1°C). Only 9" of precipitation falls statewide on the average, with areas east of the Sierra Nevada receiving as few as 4" a year.

People & History

Cave dwellers lived in the future Nevada beginning at least 20,000 years ago. Indian people were here by the 1800s and included Shoshone, Paiute, Mojave, and Washoe. They were met by Spanish explorers, then missionaries, beginning in 1775, and then traders from Los Angeles and Santa Fe in the 1830s. English trappers arrived in 1825, and Americans two years later. By the 1840s, Americans were traveling to Donner Pass in the Sierra Nevada; the trickle became a flood when gold was found in California in 1849.

Nevada had been a distant part of Spanish claims that went to Mexico when it won its independence from Spain in 1821, then to the U.S. after the Mexican War ended twenty-seven years later. In 1849, most of Nevada became part of Deseret (see Utah). Serious settlement began here when the Comstock Lode (silver and gold) was found in 1859. Nevada Territory was established in 1861.

Wanting its riches and Union sympathizers, Congress accepted Nevada as a state during the Civil War in 1864, even though it had only one-fifth of the required population. Mining—and ranching to feed miners—controlled Nevada's economy until the 1870s. After the U.S. government made silver less important for currency in 1873, ranching dominated, but profits varied with beef price changes, railroad freight rates, and frequent droughts.

In the early 20th century, more silver discoveries and competing railroads made mining profitable again. Irrigated farming began and ranchers raising their own hay increased their herd sizes. World War I increased demands for metals and beef, but the 1930s Great Depression led Nevada to diversify its economy. When gambling was legalized in 1931, resort development began.

Hoover Dam's completion in 1936 decreased electricity costs, encouraging manufacturing. During World War II, Nevada's uranium deposits suddenly became important. In the 1950s, nuclear research facilities were built.

Industry

Service industries account for one-third of Nevada's economy, higher than in any other state, and they are fed by 40 million visitors per year. Nevada mines about two-thirds of the nation's gold, and more silver than any other state. Agriculture profits mostly from cattle, sheep, and hay. Manufactured products include printed items, concrete, food products, and neon signs.

Visiting Nevada

- Nevada is the only state that allows gambling everywhere, from major casinos to slot machines in stores.

- Las Vegas has theme gambling casinos and hotels, nightlife, major entertainers, an 1850s Old Las Vegas Mormon Fort, Southern Nevada Zoological Park, Bonnie Springs Old Nevada with Old West exhibits and performances. Seventeen miles away are Red Rock Canyon National Conservation Area's sandstone formations and canyons.

- Great Basin National Park near the Utah border includes tours of Lehman Caves, a scenic drive, and backcountry hiking among alpine lakes.

- Lake Mead National Recreation Area on the Arizona border holds one of the world's largest reservoirs, backed up from Hoover Dam, one of the largest concrete dams on the Colorado River.

- Virginia City is the town the Comstock Lode built; today it has a living history museum with exhibits, tours of mansions and Piper's Opera House, and rides on the Virginia & Truckee Railroad to Gold Hill.

- Incline Village on Lake Tahoe is a four-seasons resort with golf, skiing, sandy beach, hiking trails, boating, and the Ponderosa Ranch and Western Theme Park with the set from television's Bonanza.

- Elko is a center for summer and winter outdoor recreation, and it is home to Northeastern Nevada Museum with art and historical exhibits, films, and horse-drawn vehicles.

- Reno is a smaller version of Las Vegas casinos and nightlife near Mt. Rose Ski Resort, with arboretum, botanical gardens, Fleischmann Planetarium, and mining museum.

- The Silver State's capitol in Carson City has a silver dome. Around town are the Nevada State Museum in the Old Mint Building, Children's Museum with walk-in kaleidoscope, railroad museum, and firefighter museum.

NEW HAMPSHIRE

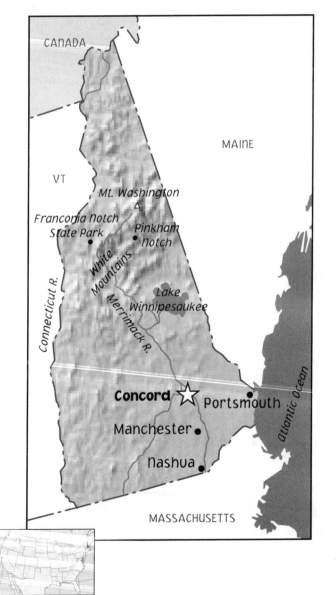

CANADA

VT

MAINE

Mt. Washington

Franconia Notch
State Park

Pinkham
Notch

White
Mountains

Connecticut R.

Merrimack R.

Lake
Winnipesaukee

Concord ☆
Portsmouth

Atlantic Ocean

Manchester

Nashua

MASSACHUSETTS

square miles: 8,968 (44th largest)
population: 1,235,786
density: 138 people per square mile
capital: Concord
largest city: Manchester
statehood: June 21, 1788 (9th state)
nickname: The Granite State
motto: Live Free or Die
bird: purple finch
flower: purple lilac
tree: white birch

Land

New Hampshire touches the Atlantic Ocean with only 13 miles of coastline (the fewest of any coastal state). The state's highest point, Mount Washington in the White Mountains (6,288'), is about 90 air miles from the coast. Most of southern New Hampshire is eastern New England uplands of hills, river valleys of the Connecticut and Merrimack, lakes, and rich soil. The Connecticut River forms the western border with Vermont. Four of the Isles of Shoals, nine miles out in the Atlantic, are part of the state.

Climate

New Hampshire enjoys cool summers but also has cold, snowy winters. The average January temperature is 19°F (-7°C), and the July average is 68°F (20°C). Annual precipitation is 42" of moisture, including snowfall of 50" near the coast and 100" in the north and west.

People & History

Indians of the Pennacook Confederacy tolerated the first European settlers in future New Hampshire in 1623, but later turned against them and joined King Philip's War (1675–76). After defeat, the Indians moved into Canada. The English colony of New Hampshire, which included today's Vermont, was governed through Massachusetts until 1679. Disputes with Massachusetts and New York over what became Vermont continued through the whole Colonial period.

Beginning in 1719, English settlers were joined by Scotch-Irish immigrants. A self-sufficient culture developed, leading to great support for the American Revolution. Locals captured a British fort at New Castle in 1774, taking its weapons and ammunition. New Hampshire was the first colony to adopt its own constitution, in January 1775, and declared its independence from Britain weeks before the United States did in 1776. Many men joined the Continental Army and the local militias during the Revolutionary War. When the state was the ninth one to ratify the U.S. Constitution, in June 1788, the Constitution was officially accepted.

With its agricultural base, New Hampshire prospered in the new nation's early years. Its swift rivers were then harnessed to run factories early in the Industrial Revolution. Manchester received its first textile mill in 1805 and soon became a manufacturing center. As factories were built after the Civil War, French Canadians immigrated to work in them, becoming the largest non-English ethnic group in the state. Portsmouth, at the mouth of the Piscataqua River on the Atlantic coast, became a shipbuilding center and port city.

After 1920, woolen and cotton mills declined as companies moved out of old factories and relocated in southern states. Shoemaking, which had been a major employer, also declined. Textile production increased during World War II, and the Portsmouth Naval Shipyard (in Maine across from Portsmouth) repaired warships and built submarines.

Industry

Service industries are the largest segment of New Hampshire's economy, including finance, real estate, insurance, and tourism. With no state sales tax, New Hampshire's retail trade attracts out-of-state shoppers and features many outlet stores. Many commuters employed in Massachusetts live in southern New Hampshire because the state has no income tax. Manufactured products include computers, electronic goods, scientific instruments, and machine tools. Only 9% of the land is given to agriculture today, mainly dairy farms; hay for cattle feed is the main crop. Maple syrup production and Christmas-tree farms add to the agricultural segment. With 85% of the state forested, there are significant logging and timber products and pulp and paper production. Granite of various colors is an important natural resource, giving the state its nickname.

Visiting New Hampshire

- Ski resorts scattered through the White Mountains attract winter athletes and spectators at major competitions. In the summer, hiking and camping are the features, and autumn's brilliant colors attract crowds of leaf peepers.

- Hampton Beach on the Atlantic Ocean is considered by many as one of the east coast's best beaches.

- Saint-Gaudens National Historic Site allows tours of the home and studio of Augustus Saint-Gaudens, 19th-century sculptor of heroic bronze statues.

- Portsmouth offers Strawberry Banke, a 10-acre restored Colonial seaport village with exhibits, and 40 buildings, the earliest dating from 1695.

- Mount Washington's cog railway takes visitors to its peak; hiking trails and a scenic drive are additional ways to explore.

- Lake Winnipesaukee, at 70 square miles, is New Hampshire's largest; boating, fishing, and water sports are allowed.

- Profile Mountain, part of Cannon Mountain, held what was the 40'-tall Old Man of the Mountain, a rock formation resembling a man's profile, which is the state's official symbol. Sadly, the face of the old man fell in May of 2003 due to natural causes.

- Cathedral of the Pines, at Rindge, is a nondenominational outdoor memorial to those who died in U.S. wars.

NEW JERSEY

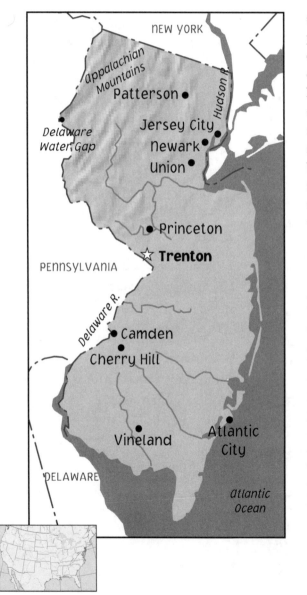

square miles: 7,417 (46th largest)
population: 8,414,350
density: 1,134 people per square mile
capital: Trenton
largest city: Newark
statehood: December 18, 1787 (3rd state)
nickname: The Garden State
motto: Liberty and Prosperity
bird: Eastern goldfinch
flower: purple violet
tree: red oak

Land

New Jersey rises from its 130 miles of Atlantic and Delaware Bay coast to the Appalachian Mountains, where High Point is just that at 1,803'. The Palisades—200' to 540' cliffs on the Hudson River—are protected as Palisades Interstate Park. The coastal plain in the south covers nearly half the state's area. The central plain, 20% of the land, is home to 70% of the population, with contiguous cities, highways, and roads. In the Highlands, narrow flat-topped ridges hold small lake-filled valleys. The Delaware River forms the Pennsylvania border, with southwestern New Jersey edging the Delaware Bay, and the Hudson River the New York border.

Climate

Influenced by westerly winds, New Jersey's climate is more inland than coastal. January sees average temperatures of 26°–37°F (-2°C–3°C), and July 70°–76°F (21–24°C). Precipitation averages 40" on the coast and 50" at higher elevations. Late winter can see severe northeasterly storms, and warm-month hurricanes occur occasionally.

People & History

As future generations would, the Delaware Indians summered on New Jersey's beaches, enjoying the fruits of the sea. French-employed Italian explorer Giovanni da Verrazano met them there in the summer of 1524, but white settlement did not begin until after Englishman Henry Hudson explored Newark Bay in 1609. Swedish, Dutch, English, and Scotch-Irish settlers came, and an English colony was established in 1664. The Dutch disputed this claim until 1669. New Jersey shared a Colonial governor with New York until 1738.

New Jersey's colonists always did their best to ignore Colonial governors, yet many residents were Tories (loyal to Britain) and did not want independence. Nearly 100 Revolutionary War battles occurred here, two of which (Yankee victories) in New Jersey gave new energy to George Washington's poorly supplied troops: at Trenton on December 26, 1776, and at Princeton over British commander Charles Cornwallis a week later.

Shortly after the war, in 1791, Alexander Hamilton worked to build an industrial city at Paterson. Although he failed, he seems to have seen part of the state's future. Canal, then railroad, building in the 19th century gave New Jersey an early lead in transport. As manufacturing grew, many truck (market produce) farms also were developed, and New Jersey became a rich state.

Although New Jersey was a Union state, residents tolerated Southern sympathizers in their midst. America's great 19th century waves of immigrants broadened the state's ethnic mix, including European and Russian Jews, Germans, Slavs, Italians, and Irish. Many African Americans left the impoverished south for jobs in New Jersey, especially around and after World War II. More recently, Puerto Ricans and Cubans have settled here. By 1970, 60% of the state's residents lived within 30 miles of New York City, where many commuted to work.

Industry

Service industries mean more to New Jersey's economy than manufacturing today, although the state is a manufacturing leader. Services span finance, insurance, and tourism. Chemicals and pharmaceuticals are the largest manufactured products, followed by food processing. Tourism boomed in the 19th century with Atlantic beach summer resorts; today summer resorts and hotels continue to prosper along with casinos in Atlantic City. Scientific and technological research became major industries during the 20th century, beginning with Thomas Edison and Samuel F. B. Morse. The state is a U.S. leader in containerized freight, which arrives on ocean-going vessels and is then transferred to trucks and airplanes. Greenhouse and nursery products (including shrubs, poinsettias, and millions of roses) earn the most agricultural income, followed by milk. Coastal commercial fishing's primary catch is clams.

Visiting New Jersey

- The Delaware Water Gap is 2.5 miles long, with walls 1,400' high, where the Delaware River flows through the Kittatinny Mountains. Protected in a National Recreation Area, it offers camping, hiking, and river swimming.

- In Atlantic City, America's first beach boardwalk was built in 1870. Today casinos and night life, along with beach and boardwalk, make it a major East Coast destination.

- Liberty State Park overlooks the Statue of Liberty and Ellis Island, an immigrant reception center from 1892 to 1954, which is restored to its 1920 appearance.

- Trenton served as the U.S. capital in 1784 and again in 1799. Visitors today can see the gilt-domed New Jersey Capitol; nearby New Jersey Cultural Center includes the state library, state museum, and a planetarium.

- Edison National Historic Site, West Orange, shows Thomas Edison's home and laboratories, with examples of his many inventions, including phonograph and motion picture apparatus.

- Cape May, settled in 1631 on the southern tip, was America's first summer ocean resort. Today's visitor can view 250 Victorian buildings, including "cottages" of mansion proportions.

- Stanhope's Waterloo Village recreates Colonial life in the 1700s on the Morris Canal, with furnished homes, blacksmith shops, and mill.

- Morristown National Historic Park holds soldiers' camps of the Revolutionary War, restored Fort Nonsense (dating from 1777), and Ford Mansion, Washington's headquarters for winter 1779–80.

NEW MEXICO

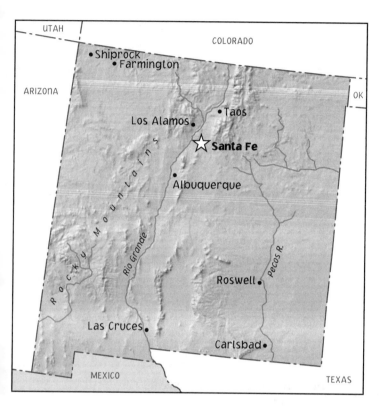

square miles: 121,356 (5th largest)
population: 1,819,046
density: 15 people per square mile
capital: Santa Fe
largest city: Albuquerque
statehood: January 6, 1912 (47th state)
nickname: The Land of Enchantment
motto: Crescit Eundo (It Grows as It Goes)
bird: roadrunner
flower: yucca
tree: piñon

Land

New Mexico's eastern third is Great Plains, its central third rises in the Rocky Mountains—including Sangre de Cristo range where Wheeler Peak, 13,161', is the state's highest point—and the western third's high plateaus are broken by short mountain ranges. Red Bluff Reservoir on the Texas border, at 2,817', is the state's lowest point. The Rio Grande and Pecos rivers flow south, the Gila and San Juan flow west.

Climate

January statewide temperatures average 34°F (1°C), and July averages 74°F (23°C), with local temperatures affected by elevation. Statewide annual average precipitation of 13" includes more in the east and far less in west, plus mountain snow. Surface water is scarce, but New Mexico holds underground aquifers, which serve as natural reservoirs.

People & History

Ancient people lived along streambanks 10,000 years ago in future New Mexico. Mogollon Indian villages rose near the Arizona border from 500-1200 A.D. Anasazi built their apartment buildings, some carved into cliffs, in the northwest. They raised cotton, squash, corn, and turkeys, making clothing of feathers. In the 16th century, Navajos, Apaches, Utes, and Comanches moved here.

Spanish explorers, hearing of seven cities of gold, searched New Mexico for them in 1539 and 1540–42. Fifty-six years later, the first Spanish colony was founded. Its capital moved to Santa Fe around 1610, making that city the oldest U.S. seat of government.

Catholic priests established missions where they forced the Indians to farm. In 1680, led by a Pueblo called Popé, the Indians drove their masters into Mexico and destroyed the missions. Spain reconquered the land in 1692, and after four years of Indian raids, peace returned.

Mexico won independence from Spain in 1821, with New Mexico as a province. Indians and New Mexicans rebelled but lost in 1837, and Texas invaded in 1841 but was driven out by Mexico. Most of New Mexico came under U.S. control after the Mexican War ended in 1848, and the Gadsden Purchase of 1853 obtained its southern portion. Today, 40% of New Mexicans are Hispanic.

Early in the Civil War, Texas Confederates captured much of the state, which was retaken by Union troops in 1862. From 1862 to 1864, the army battled Navajo and Mescalero Apaches until forcing them onto reservations. Other Apaches fought settlers until 1886, when their leader Geronimo surrendered.

Trade with the U.S. opened in 1821 when the Santa Fe Trail stemmed from Missouri, but arrival of railroads beginning in 1879 fostered a mining and cattle boom. Severe drought in the 1920s saw many cattle ranches (and banks) fail, but oil was discovered at the same time.

In Los Alamos during World War II, the military secretly researched and manufactered the atomic bomb. The weapon was first exploded near Alamogordo in July 1945, just a month before being dropped on Hiroshima, Japan.

Industry

Federal government employment—in national parks, on Indian reservations, and in nuclear research facilities—and tourism forms the largest portion of New Mexico's economy. Reserves of uranium, copper, coal, oil, natural gas, and potash (used in fertilizer) make this an important mining state. Cattle ranching is the primary agricultural activity, and electronics the main manufactured item.

Visiting New Mexico

- Visitors to Carlsbad Caverns National Park can take a three-mile walk through some of the largest underground caves in the world, where the temperature is always 56°F.

- Petroglyph National Monument holds 15,000 examples of prehistoric and historic Native American rock carvings and five volcanic cones.

- Seventeen Indian pueblos and three reservations welcome visitors.

- Albuquerque's Sandia Peak Tramway moves sightseers to an observation deck and cafe at 10,378'. Old Town around the plaza includes San Felipe de Neri Church (built 1706), an Indian Pueblo Cultural Center, restaurants, and stores. The city is home to the Children's Museum and National Atomic Museum, among others.

- Aztec Ruins National Monument, though misnamed, allows self-guided tours of Anasazi pueblo's homes and ceremonial buildings.

- White Sands National Monument protects dunes of gypsum, whiter than most sand, and allows hiking and dune surfing.

- Taos mixes art colony, ski resort, whitewater rafting, horseback rides, llama treks, and Taos Pueblo, northern New Mexico's largest.

- Santa Fe's central plaza was the end of the Santa Fe Trail; on it, Palace of the Governors, dating from 1610, is the oldest public building in the U.S. Several museums exhibit art and Indian cultures and arts. The town hosts 250 art galleries, 200 restaurants, and in July and August, the Santa Fe Opera.

NEW YORK

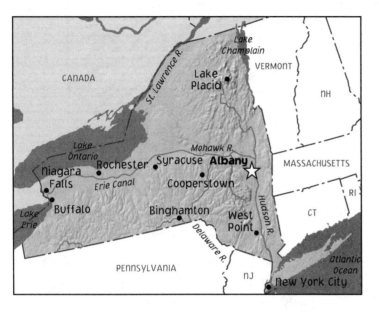

square miles: 47,214 (30th largest)
population: 18,976,457
density: 402 people per square mile
capital: Albany
largest city: New York City
statehood: July 26, 1788 (11th state)
nickname: The Empire State
motto: *Excelsior* (Ever Upward)
bird: bluebird
flower: rose
tree: sugar maple

Land

New York touches water on its east and west: 127 miles of Atlantic Ocean coastline and 371 miles of Great Lakes coast on Lakes Erie and Ontario. Its northern border is the St. Lawrence River. Additional lowlands are the Mohawk and Hudson River valleys in eastern New York. In the northeast, the Adirondack Mountains include 5,344' Mount Marcy, the state's highest point. Half the state is Appalachian highlands; this includes the Catskill Mountains, Finger Lakes hills, and Delaware River basin.

Climate

Statewide, January's average temperature is 21°F (-6°C), with higher averages on Long Island Sound and lower ones in the mountains. July's statewide average temperature is 69°F (21°C). Statewide annual precipitation of 39" includes large amounts of snow in the mountains and land east of the Great Lakes. Few New York State days see completely clear skies.

People & History

Future New York was home to Mohican and Iroquois Indians before Europeans arrived. Henry Hudson, an Englishman working for Holland, sailed up the Hudson River in 1609, and the Dutch built New York's first settlement, Fort Orange, at today's Albany fifteen years later. In 1625, they created New Amsterdam on the southern end of Manhattan Island, in today's New York City. Life in Holland was comfortable, and few Dutch settlers came to the New World. The Dutch gave up their colonies to England in 1664.

New York colony grew slowly, because it was a battleground for the French and Indian War between France and England. Mohicans sided with France, and Iroquois with England in fighting that lasted until 1763. Britain and the Iroquois won, and settlers began to move into New York. As the American Revolution began, though, New York ranked seventh in population among colonies.

More people here than in any other colony remained loyal to England, and in many American Revolution battles in New York neighbors fought each other. Britain invaded from the east and north, and one-third of Revolutionary War battles occurred in New York.

After the war, New York grew quickly, and its main port, New York City, became a center of commerce and transportation. By 1810, it was the most populous state, a rank held until about 1963. Industrial development in the 19th and 20th centuries centered on Buffalo, Rochester, Albany-Troy-Schenectady, Syracuse, and Binghamton, during which time New York City became a financial center and home of the garment industry.

During the 19th century, New York grew with immigrants from around the world who entered through the port of New York City. As the nation's largest city, New York City has a huge influence on its state.

Industry

Services are the largest segment of New York's economy, including finance, insurance, real estate, advertising, entertainment, law, and wholesale and retail trade. Publishing and printing are important industries along with cameras, copiers, optical and dental instruments. One-fourth of the state is still farmland, and milk is the major product, followed by beef, hay, and corn. Commercial fishing exists on Long Island Sound and Lakes Erie and Ontario.

Visiting New York

- New York City served as the U.S. capital 1788–90 and today hosts the United Nations headquarters. It is the largest U.S. city and the world's fifth largest, the nation's financial and cultural center, and it offers a wide variety of performance and visual arts, museums, historical sites, ethnic neighborhoods, professional sports teams; the Statue of Liberty erected 1886 in the harbor; and the Empire State Building, a skyscraper completed in 1931.

- Cooperstown's National Baseball Hall of Fame and Library attracts many visitors; in addition, Fenimore House offers American folk art, Farmers' Museum shows agricultural history, and Village Crossroads presents businesses and homes from the 1700s and 1800s.

- At Niagara Falls, water flows at 100,000 cubic feet per second over American Falls (in New York) and Horseshoe Falls (in Canada), which are surrounded by parkland.

- Britain's stone mountaintop Fort Ticonderoga, near the village of the same name, fell to Vermonters in 1775 as the movement for independence began; visitors today see a reconstruction with costumed interpreters.

- In Corning, the Corning Glass Center demonstrates glassmaking and presents its history in the Museum of Glass and Hall of Science and Industry.

- Empire State Plaza in Albany borders state government buildings and leads to cultural facilities including the state historical museum.

- Long Island Sound, the Finger Lakes, Catskill Mountains, and Adirondack Mountains are popular summer resorts. Lake Placid in the Adirondacks, site of the 1980 Winter Olympics, hosts winter sports and competitions.

- The St. Lawrence River's Thousand Islands number more like 1,700, and can be enjoyed on the water from resorts or viewed from the multipart, 6.5-mile Thousand Islands International Bridge.

NORTH CAROLINA

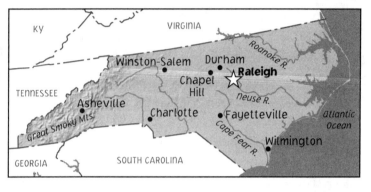

square miles: 48,711 (28th largest)
population: 8,049,313
density: 165 people per square mile
capital: Raleigh
largest city: Charlotte
statehood: November 21,1789 (12th state)
nickname: The Tar Heel State
motto: *Esse Quam Videri* (To Be Rather Than to Seem)
bird: cardinal
flower: dogwood
tree: pine

Land

The eastern half of North Carolina is low-land—average elevation 20'—from the swampy Tidewater along the Atlantic to the rolling coastal plain. Three capes—Hatteras, Lookout, and Fear—reach into the Atlantic, creating dangerous waters nicknamed the Graveyard of the Atlantic in the days of sailing ships. About 130 miles inland the Piedmont (foot of the mountains) begins. These rolling, forested hills cover more than a third of the state. Westward, two ranges of the Appalachian Mountains rise: the Blue Ridge and Unaka Mountains that include the Great

111

Smoky Mountains. The Blue Ridge's Mount Mitchell, at 6,684', is the highest point in the eastern U.S.

Climate

North Carolina's July average temperature is 70°F (2°C), with January's average 41°F (15°C). Statewide average annual precipitation is 50". July and August are the mountains' rainiest months, and October–November their driest. The coast is second only to Florida's in number of hurricanes.

People & History

Although both France and Spain explored the coast, only Britain colonized it. Sir Walter Raleigh sent two expeditions, followed by colonists, to Roanoke Island in the Outer Banks in 1584, 1585, and 1587. The last group established Britain's first North American colony. Secotan Indians first welcomed then fought the newcomers, then suffered from smallpox that the settlers brought, and then loss of land. In 1590, returning Englishmen found no sign of the settlers. The "Lost Colony's" fate is unknown.

In the 1650s English colonists tried again, moving into the Piedmont from Virginia and Pennsylvania. The colony's white population included English, Irish, Scottish, Welsh, French, German, and Swiss. Fighting enslavement, Tuscarora Indians battled colonists, then moved out in the early 1700s to join Iroquoian relatives. African American slaves brought into the colony formed one-third of the population by 1790.

Two cultures developed. In the east, large cotton and tobacco plantations used slave labor; in the west, small farms without slaves were the way of life. The American Revolution temporarily united the colony.

In the 1830s, when the U.S. government declared that Indians must move west to today's Oklahoma, a few hundred Cherokees escaped into the mountains and avoided the "Trail of Tears" six-month march. Today their descendants, the Eastern Cherokees, have a reservation in the Smokies.

North Carolinians remained divided over slavery, and many citizens were against leaving the Union when the state did so in 1861. The war didn't harm North Carolina's important agricultural lands, and its economy, subsequently rose and diversified. Charlotte became both Carolinas' largest city area, a banking and service center.

Industry

Agriculture has become agribusiness, the main products being tobacco, broiler chickens, turkeys, and hogs. Furniture factories, wood and paper products, and textile mills moved North Carolina into the top third of U.S. manufacturing states. Forests are managed for regrowth, which is helped by the state's mild climate. Today's Research Triangle—bounded by Duke University, the University of North Carolina, and North Carolina State—is a center for government and private scientific research facilities. Freight reaching deepwater port Wilmington is only one day by truck from New York or Florida, making transport a major industry.

Visiting North Carolina

- Great Smoky Mountains National Park straddles the Tennessee border, with half its 520,000 acres in North Carolina. It includes walking and horse trails, waterfalls, lush forest, trout fishing, and exhibits on Cherokee history and on mountain farms.

- Biltmore Estate near Asheville, set in 7,500 acres of landscaped and natural land, centers on the 250-room French Renaissance mansion of a Vanderbilt grandson.

- New Bern, settled by Swiss and Germans, was both the Colonial and first state capital. Maps outline a self-guided tour, and Tryon Palace Restoration and Gardens Complex has the government building that was envied by the other twelve colonies.

- The Outer Banks, islands along 125 miles of the coast, hold primitive areas and developed resorts. Kill Devil Hills is where the Wright Brothers flew their first airplane in 1903, and Cape Hatteras National Seashore the longest stretch of undeveloped seashore on the U.S. Atlantic.

- Bath, the first incorporated town in 1705, was the colony's first seaport and home port of Blackbeard the pirate.

- Tweetsie Railroad, north of Blowing Rock, is a restored narrow-gauge coal-fired train. The surrounding theme park includes trains, an 1880s railroad town, crafts demonstrations, and live music.

- Blue Ridge Parkway travels the top of the mountains, up to 6,053', for 469 miles between Great Smoky Mountains National Park and Shenandoah National Park (Virginia). Many visitor centers and exhibits are along the way.

- Beaufort's North Carolina Maritime Museum covers coastal natural history and human history.

- At Old Salem in Winston-Salem, costumed guides explain life in the original Moravian settlement here in 1766.

NORTH DAKOTA

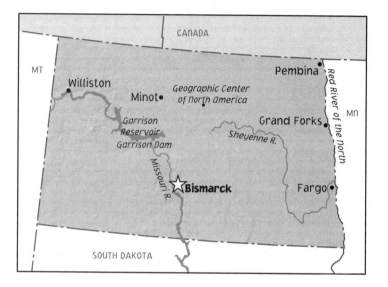

square miles: 68,976 (17th largest)
population: 642,200
density: 9 people per square mile
capital: Bismarck
largest city: Fargo
statehood: November 2, 1889 (39th state)
nickname: The Sioux State, The Flickertail State
motto: Liberty and Union, Now and Forever, One and Inseparable
bird: Western meadowlark
flower: wild prairie rose
tree: American elm

Land

North Dakota's eastern half is part of the central lowland, with the Red River valley (its eastern border) the state's lowest point, at 750'. From its center, the state rises to the Drift Prairie and then the Great Plains. In the southwest, the Badlands hold White Butte, North Dakota's highest point at 3,506'. The Missouri River flows in from the west and turns south in west-central North Dakota.

Climate

Although North Dakota's cool, dry climate sees some very hot summer days, July's average temperature is 70°F (21°C); January averages 7°F (-14°C). Average annual precipitation is 17".

People & History

Future North Dakota was home to Sioux Indians, to Mandan, Hidatsa, and Arikara in the Missouri River valley, and to Chippewas and Crees in the northeast. Sieur de la Vérendrye, a Canadian, was the first recorded white visitor in 1738. Canadian fur traders traveled here regularly by the 1790s.

After this land came under U.S. control in the Louisiana Purchase, the Lewis and Clark Expedition spent its first winter—1804–5—with the Mandans and Hidatsas near today's Bismarck.

American traders began traveling the Missouri in the 1820s and 1830s, bringing native people useful tools but also diseases for which they had no immunity. River traffic increased greatly when gold was discovered in Montana (1862) and Indian hostility grew. The U.S. army built forts along the rivers to protect travelers.

With the first railroad arriving in 1871, settlers—mostly from northern Europe—claimed land under the Homestead Act and established wheat farms. In 1890, North Dakota had the highest percentage (not quantity) of foreign-born residents of any state.

Dependent on railroads to move crops to market, farmers came to resent the economic control these corporations had. Movements began shortly before World War I that resulted in establishing a state-owned bank, flour mill, and grain elevator, and cooperatives for selling grain and purchasing farm supplies.

Recent years have seen greater farm mechanization with an increase in average farm size. Oil discovered in western North Dakota in the 1950s added an industrial component to the state's economy. Airbases and missile silos were built here in the 1960s.

Industry

Agriculture dominates the economy of North Dakota, which usually leads the nation in growing high-protein wheat; other grains, sunflower seeds, and sugar beets are important crops. Cattle, hogs, sheep, and turkeys are raised. Part of the Williston Basin oilfield is in North Dakota, and the state has large deposits of lignite coal. Much of the manufacturing sector is devoted to processing food, oil, and coal. The state exports electrical power produced in lignite-fired plants and at hydroelectric Garrison Dam on the Missouri.

Visiting North Dakota

- Bismarck is marked by the 19-story-skyscraper state capitol rising above city and surrounding plains, completed in 1932 with Art Deco motifs; an observation deck tops it. On the capitol grounds, North Dakota Heritage Center's exhibits extend from prehistoric peoples to the present. Riverboat Lewis & Clark cruises the Missouri River daily. Camp Hancock Museum commemorates the importance of railroads in North Dakota development.

- Fort Lincoln State Park at Mandan includes reconstructed buildings of the fort from which Colonel George Custer led part of the U.S. Seventh Cavalry to defeat at the Battle of the Little Bighorn (Montana) in 1876. Also in the park is On-A-Slant Indian Village, where costumed guides and working crafters show Indian farming life of the 1600s.

- Washburn's Lewis and Clark Interpretive Center tells of the expedition's entire trip, with emphasis on the winter of 1804–5 spent here.

- Knife River Indian Villages National Historic Site at Stanton has excavations of three Mandan earthlodge villages the Lewis and Clark Expedition visited, where they met their future interpreter, Shoshone teenager Sacagawea (spelled Hidatsa-style "Sakakawea" in North Dakota).

- Theodore Roosevelt National Park in the Badlands is where the future president ranched from 1883–86. Guided horseback rides, scenic drives, and hiking take visitors among wildlife from buffalo to prairie dogs, to Roosevelt's ranch site, and to the 26-room mansion of French-born rancher Marquis de Mores.

- Fort Union Trading Post National Historic Site near the Montana border reconstructs an important fur-trading post built by John Jacob Astor's company, which functioned from 1828 to 1867.

- International Peace Garden in North Dakota and Manitoba has gardens, outdoor recreation, and summer open-air concerts celebrating the U.S.-Canada border, the world's longest unfortified international boundary.

OHIO

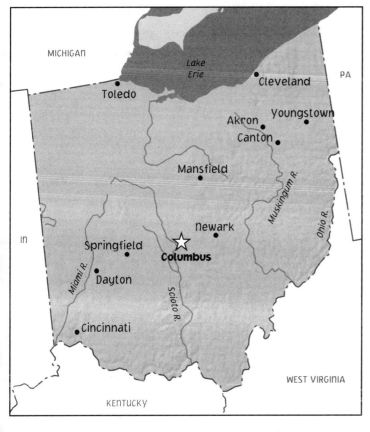

square miles: 40,948 (35th largest)
population: 11,353,140
density: 277 people per square mile
capital: Columbus
largest city: Columbus
statehood: March 1, 1803 (17th state)
nickname: Buckeye State
motto: With God, All Things Are Possible
bird: cardinal
flower: scarlet carnation
tree: buckeye

Land

Ohio curves 226 miles along Lake Erie, its southern edge roughly parallel to the Ohio River. Northern Ohio holds Lake Plains' level to rolling hills. Ohio's eastern half includes rolling hills of the Allegheny Plateau, steeper in the southeast. Western Ohio's fertile soil is the easternmost reach of the Central Plains. Campbell Hill, near Bellefontaine, is the highest point at 1,550'; Ohio's lowest point is 433' where the Miami and Ohio rivers confluence.

Climate

Ohio's humid climate sees rainstorms (and occasional tornadoes) in summer and snowstorms in winter, with statewide average precipitation averaging 38". July temperatures average 73°F (23°C) while January averages 28°F (-2°C).

People & History

People of mound-building Hopewell and Adena cultures lived in future Ohio until about 1500, creating earthwork forts and burials. They were followed by Miamis, Shawnees, Wyandots, and Delawares, among other Indian tribes.

French exploration along Lake Erie and the Ohio River led that nation to claim the area in the late 1600s. Britain and France fought over it in the last French and Indian War beginning in 1754. In 1788, New Englanders founded the future state's first permanent town, Marietta.

Indian raids were frequent until the Battle of Fallen Timbers in 1794 and the resulting Treaty of Greenville. During the War of 1812, the American navy beat Britain's in the Battle of Lake Erie, late 1813.

Population grew quickly with the rich farmland and also the transportation corridor between the east coast and the frontier. Lake Erie shipping, Ohio River barges, and—beginning 1838—National Road freighting carried supplies westward. Now part of U.S. Route 40, the National Road to Illinois was the first federal highway. Two main canal systems, built in the 1830s and the 1840s, were important until railroads arrived in the 1850s.

Mainly German farmers and Irish canal builders followed the original New England settlers. Later in the 19th century, factories attracted immigrants from all parts of Europe and African Americans from the South.

After the Civil War, Ohio became a leading industrial state. It also was a national political force, with seven Ohio-born U.S. presidents elected from 1840 to 1920. By the 1980s, outmoded factories and foreign competition cut manufacturing, but new factories have since been added. After 1900, self-sufficient farms began to give way to today's specialized ones.

Industry

One-fifth of Ohio workers are employed in manufacturing, 30% of the state's economy. They make automobiles, rubber products, machine tools, porcelain, pottery, and steam shovels. Mining contributes coal, limestone, and salt from the nation's deepest salt mine. Oceangoing vessels dock at Toledo and Cleveland, arriving via the St. Lawrence Seaway. Corn and soybeans are the most important farm products, followed by cattle and hogs. Service industries, including tourism, make up two-thirds of the economy.

Visiting Ohio

- Great Serpent Mound, Hillsboro, was built by ancient people for unknown reasons. A quarter-mile long, the earthwork forms the curved figure of a snake swallowing an egg.

- Sixteen covered bridges are in Ashtabula County, and Hubbard House, an Underground Railway "station" for escaping slaves, is in the town of Ashtabula.

- Schoenbrunn Village, founded 1772 by Moravian missionaries, is reconstructed near New Philadelphia, and has Ohio's first school-house.

- Dayton, home of the flying Wright brothers, exhibits their 1905 Wright Flyer in Carillon Historical Park, and Wright-Patterson Air Force Base has the U.S. Air Force Museum. Dayton Art Institute hangs permanent and changing exhibitions. Carriage Hill Farm brings 1880s farming to life.

- Cleveland has the Rock and Roll Hall of Fame and Cleveland Museum of Art, Frederick C. Crawford Auto-Aviation Museum, and tours of NASA Lewis Research Center. The Cleveland Symphony moves outdoors to Blossom Music Center during summer. Nearby Bath features Hale Farm and Village, dating from the 1820s.

- Canton is home to the Pro Football Hall of Fame, with films, videos, uniforms.

- Greenville's Garst Museum exhibits collections from native sharpshooter Annie Oakley (often advertised as a "Girl of the West") and newsman Lowell Thomas.

- Coshocton's Roscoe Village puts visitors in an 1830s setting on the Ohio and Erie Canal, with slow, quiet rides on canal boats pulled by horses walking the bank.

- German Village in Columbus is a restored mid-1800s neighborhood with shops, festivals, and ethnic foods. At Ohio Historical Center, Ohio Village's 14 buildings reconstruct an 1850s county seat; indoor exhibits begin with ancient people. Parklike Columbus Zoo houses Discovery Reef's saltwater aquarium.

- From Catawba Point, ferries take visitors to South Bass Island and Perry's Victory and International Peace Memorial. The 352' monument, topped by an open walkway, honors the Battle of Lake Erie victory.

OKLAHOMA

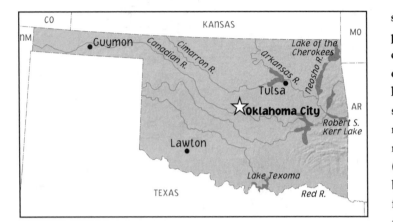

square miles: 68,667 (18th largest)
population: 3,450,654
density: 50 people per square mile
capital: Oklahoma City
largest city: Oklahoma City
statehood: November 16, 1907 (46th state)
nickname: The Sooner State
motto: *Labor Omnia Vincit*
(Labor Conquers All Things)
bird: scissor-tailed flycatcher
flower: mistletoe
tree: redbud

Land

Eastern Oklahoma is forested rolling hills, but the west features high plains. The Oachita (wash-i-taw), Wichita, and Arbuckle mountains rise in the south, and the Ozarks extend into the northeast. Oklahoma's east-to-west incline toward the High Plains makes its highest point 4,973' at Black Mesa in the northwest. Oklahoma's major river is the short, winding Arkansas of the northeast, which connects through the McClellan-Kerr Navigation System to the Mississippi, then to the Gulf of Mexico. The Red River forms the Texas border.

Climate

Oklahoma sits where dry northern continental winds meet humid southern winds, so wind is constant and tornadoes, thunderstorms, and blizzards occur. Average January temperature is 37°F (3°C), the July average is 82°F (28°C). Fifty inches of rain falls per year in the Ouachita Mountains, but less than 15' in the panhandle.

People & History

The Clovis and Folsom people lived in Oklahoma more than 10,000 years ago. Plains Indians moved in and met Coronado's Spanish explorers in 1541. French fur traders arrived by 1714, and the two nations competed until Spain gave this land to France in 1800. Three years later, the United States acquired most of future Oklahoma in the Louisiana Purchase.

In 1830, the U.S. Congress established eastern Oklahoma as Indian Territory, where native peoples from the southeast would be moved. From 1817 to 1846 came the Chickasaws, Creeks, some Seminoles, Choctaws, and Cherokees—their forced march is called the Trail of Tears because many died en route. At the U.S. Civil War, many of these Indian nations, whose cultures used slavery, supported the Confederacy, but others served the Union or stayed neutral.

By 1872 the Osage were in Indian Territory, and by 1875, the northern plains Cheyenne, and southwestern Arapaho, Kiowa, and Comanche. By 1885, people from 50 tribes—including Kickapoo, Sauk and Fox, Kaw, Seneca, and Wyandot—had been moved into Oklahoma.

Today, although reservations remain, no other state has greater integration of Indian residents (about 8% of the population) into its general population. White settlers (called Boomers) began moving into Indian Territory without permission in 1879. Although the federal government removed many, it opened 2 million acres of non-Indian land to settlers in April 1889. Perhaps 60,000 people participated in the land rush; those who sneaked in early were called Sooners. In 1893, in another land rush, 100,000 people settled on 6 million acres. Indian (east) and Oklahoma (west) Territories remained separate until filing for statehood jointly.

Major oil discoveries near Tulsa, 1901 and 1905, made the city Oklahoma's main oil refining center, but a large percentage of profits went out of state and agriculture remained important. In the 1930s, climate collided with farming practices, creating the Dust Bowl as winds blew away dry topsoil. Many farmers moved away in search of work.

Boom times of World War II let Oklahoma diversify its economy. Its cities grew, and by midcentury, claimed half the residents, mainly in Oklahoma City, Tulsa, and Lawton.

Industry

Oil is Oklahoma's most important product; natural gas, natural gasoline, butane, and coal production are also significant. Manufacturing includes machinery, food processing, and items made from metal, rubber, and plastics. Agriculture is centered on large specialty farms, raising mainly beef cattle along with winter wheat, hay, corn, sorghum, soybeans, and peanuts. Dense Oachita and Ozark forests offer pine, oak, hickory; pine supports a major paper mill and sawmills are numerous. Tulsa oil shipping is linked to the Mississippi River and the Gulf of Mexico via the McClellan-Kerr Arkansas River Navigation System (1971).

Visiting Oklahoma

- Tulsa's Thomas Gilcrease Institute of American History and Art exhibits American Western art; Tulsa Zoo and Living Museum includes habitat settings; the Philbrook Art Center is an Italian Renaissance villa, amidst gardens, that houses art from around the world.

- Claremore has the Will Rogers Birthplace Ranch, and the Will Rogers Memorial Museum with exhibits on Rogers's work in rodeo, writing, stage, and film.

- Oklahoma City houses the only state capitol with an oil well on its grounds. It also features the National Cowboy Hall of Fame and

Western Heritage Center with art and artifacts of Indians, white settlers, cowboys, and rodeos; Oklahoma State Museum of History; rodeos and horse shows; minor league baseball and basketball; and National Softball Hall of Fame and Museum.

- Chickasaw National Recreation Area near Sulphur in south-central Oklahoma includes woods, waterfalls, mineral springs, swimming and boating at Lake of the Arbuckles, camping, fishing, and seasonal hunting.

- Har-Ber Village near Grove is a reconstructed 19th-century village of more than 100 buildings filled with artifacts from that era and from earlier Indian cultures.

- Fort Sill Military Reservation and National Historic Landmark, near Lawton, shows an 1869 frontier military post and is site of Geronimo's grave.

- Anadarko is home to Indian City USA (museum plus re-creation of a Plains village), Southern Plains Indian Museum, and National Indian Hall of Fame.

OREGON

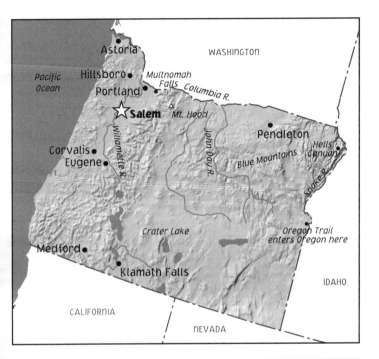

square miles: 95,997 (10th largest)
population: 3,421,399
density: 36 people per square mile
capital: Salem
largest city: Portland
statehood: February 14, 1859 (33rd state)
nickname: The Beaver State
motto: *Alis Volat Propriis*
(She Flies with Her Own Wings)
bird: Western meadowlark
flower: Oregon grape
tree: Douglas fir

Land

Oregon's 296 miles of Pacific Ocean coastline are paralleled by the Coast Range, the state's lowest mountains. South of the Coast Range are the Klamath Mountains. Inland, the Willamette Valley's rolling hills are where most people live, and here is the seat of agriculture. To their east, the Cascade Range includes Mount Hood, the state's highest point at 11,239'. The North Central Oregon Plateau has broad wheat fields cut by rivers. In the northeast, the land rises again in the Blue and Wallowa mountains. South of them, and

123

east of the Cascades, is the High Desert, which gives way to Basin and Range land in the southeast. The Columbia River forms the northern boundary.

Climate

Oregon receives 28" of precipitation per year, almost all in the form of rain. January average temperatures dip to 32°F (0°C), but July's averages rise only to 66°F (19°C).

People & History

Before the arrival of whites, Oregon's mild climate was home to up to 180,000 Indians divided among 125 tribes who lived as hunter-gatherers. Spanish explorers seeking a northwest water passage to the interior viewed the coast, but England's Francis Drake was the first white to step ashore in 1579. English captain James Cook, in 1778, traded with Indians for furs that he sold in China. Fourteen years Later, Robert Gray sailed into the mouth of the Columbia River, which he named and claimed for the U.S.

The American Fur Company built a trading post in 1811 at Astoria, named for company founder, John Jacob Astor. In the 1840s, American settlers arrived via Oregon Trail wagon trains; they farmed in the Willamette Valley. Oregon Territory was created in 1846 after Britain dropped its claims. When railroads arrived in 1883, agriculture and forestry grew with the access to national markets. The twentieth century saw related industries develop and the population grows faster than in most other states. Recent years have seen immigration of retirees and former Californians.

Industry

The timber industry and related wood-products manufacturing, such as plywood and paper, account for one-fifth of Oregon's economy. Food processing is important, but heavy industry has not developed because the state lacks iron ore and coal. Agriculture employs only 5% of the state's workers on mechanized farms that cover 10% of the land. The main crops are hay, wheat, barley, and fruits. Service industries such as government, trade, finance, and construction, employ most workers.

Visiting Oregon

- Highway 101 takes visitors 340 miles along the Pacific coast through and alongside Pacific coast beaches, resort towns, lighthouses, seafood restaurants, and whale-watching sites.

- Portland is home to Tom McCall Waterfront Park, with biking and walking trails and paddlewheeler river tours; the Rose Festival for three weeks each June includes flower shows, an Indy car race, a carnival, and performances; Portland Art Museum, Powell's (the nation's largest bookstore), the World Forestry Center, an arboretum and zoo are other offerings.

- Pendleton in eastern Oregon's ranch country has saddlemakers at work, a major rodeo in September, and Pendleton Woolen Mills. Nearby Umatilla Indian Reservation has casino gambling, dance presentations, and art exhibits.

- Crater Lake National Park, north of Klamath Falls, holds the 12,000'-deep lake formed when volcanic Mt. Mazama exploded and the hole in its top filled with water. Rim Drive runs for 33 miles around the lake, and is open June–October; tour boats follow the shoreline.

- Fort Clatsop National Memorial, at Astoria, recreates the tiny enclave where the Lewis and Clark Expedition wintered in 1805–6. Living history performances are given in summer.

- Columbia River Gorge National Scenic Area can be toured by car from Troutdale or by sternwheeler on the water. Multnomah Falls, the nation's second highest, is one of 77 here.

- Eugene and Springfield, divided by the Willamette River, are bicycle friendly and filled with concerts, exhibits, and other cultural activities centered on the University of Oregon in Eugene.

- Newport is a fishing and resort village with shops and seafood restaurants overlooking the waterfront; here, too, are the Oregon Coast Aquarium and Agate Beach, where rock and gem collectors roam.

- In Salem, the state capitol's rotunda offers views of Mount Hood and Mount Jefferson. Mission Mill Village displays early woolen production, and nearby wineries give tours.

- Mount Hood is snow-capped all summer, and its surrounding national forest holds resorts, a 170-mile loop road, fishing, campsites, and cross-country ski trails.

PENNSYLVANIA

Lake Erie
NEW YORK
Delaware R.
Erie
Bradford
Scranton
Williamsport
Susquehanna R.
Pocono Mts.
Pymatuning Reservoir
Allegheny R.
Bethlehem
Allentown
NEW JERSEY
State College
Levittown
OHIO
Ohio R.
Harrisburg
Philadelphia
Penn Hills
Hershey
Pittsburgh
Gettysburg
Monogahela R.
MARYLAND
DE
WV

square miles: 44,817 (33rd largest)
population: 12,281,054
density: 273 people per square mile
capital: Harrisburg
largest city: Philadelphia
statehood: December 12, 1787 (2nd state)
nickname: The Keystone State
motto: Virtue, Liberty, and Independence
bird: ruffed grouse
flower: mountain laurel
tree: hemlock

Land

From 40-plus miles of Lake Erie coast, the Erie Lowland extends a short distance to the Appalachian Plateau of northern and western Pennsylvania. In the latter, the Allegheny Mountains include Mount Davis, the state's highest point at 3,213'. Farther east, running southwest to northeast, the 70-mile-wide Ridge and Valley Province contains deep valleys. The Great Valley parallels it, with the rugged Blue Mountains to its east. From there, the Piedmont lowlands drop to the narrow Atlantic coastal plain. At Pittsburgh, the Allegheny and Monongahela join to form the Ohio River; the Delaware River forms Pennsylvania's eastern border, and the Susquehanna drains about half the state.

Climate

Moisture characterizes Pennsylvania's climate, with rainy, humid summers and snowy winters. Statewide, July's temperature averages 71°F (22°C), with January's averaging 27°F (-3°C); averages are slightly lower in the Erie lowland. Annual precipitation totals 41".

People & History

Delaware, Shawnee, and Susquehanna Indians lived among Pennsylvania's wooded mountains and valleys before Europeans began to arrive in 1643. Swedes were the first white settlers, near today's Chester on the Delaware River. In 1657, Dutch traders opened fur-trade posts. Fighting between Swedish and Dutch settlers led New Netherland to seize the colony in 1655. Nine years later, Holland peacefully turned over its American colonies to England. New Netherland became New York, with Pennsylvania being part of it.

In 1681, England's King Charles II chartered Pennsylvania to William Penn, a Quaker who created a colony allowing religious freedom. Under his government, Indians were treated more fairly than in other colonies, but increasing white settlement changed their once-welcoming attitude towards whites. They joined France's side in the French and Indian War, 1754–63—part of France and England's struggle for North America.

After England won that war, it banned further settlement to the west, but Pennsylvanians (especially Scotch-Irish immigrants) ignored the ruling and moved west of the Alleghenies. Fort Pitt, at the head of the Ohio River, grew as a frontier hub; it is today's Pittsburgh.

Settlement had spread from the southeast, where Penn built his capital of Philadelphia. Following English colonists came German immigrants who moved onto rich farmlands. Their word Deutsch (German) soon changed to Dutch, and they became the Pennsylvania Dutch.

Pennsylvania was a leader in the American Revolution, with the Declaration of Independence signed in Philadelphia on July 4, 1776. Following the war, Pennsylvania grew rich with agriculture, its transportation corridor to the frontier and, later, industrialization based on rich natural resources. Coal mining began in 1820; in 1859, the first successful U.S. oil well came in at Titusville.

Beginning in 1781, the state had gradually abolished slavery. By the Civil War, it was the east coast dividing line between the U.S. and the Confederacy, and halfway through that war, the three-day battle at Gettysburg ended the South's attempts to invade the North.

Irish immigrants arrived in the later 1800s, and the state's industry attracted Italians and eastern Europeans in the early 20th century. African Americans moved from the South for jobs, especially during and after World War II. Recent years have seen rises in Puerto Rican and Asian immigration.

Industry

Manufacturing is Pennsylvania's most important activity, including coal (soft and hard coal; this is the only state to supply hard coal) and oil production. Important products are steel, food, electrical and nonelectrical machinery, chemicals, and paper. In agriculture, it leads the states in producing mushrooms, apples, milk and milk products, chickens, and eggs. Overcutting of its vast forests in the 19th century led to a decline in the timber industry, now generally limited to logging for pulp and paper factories. Tourism is growing in importance.

Visiting Pennsylvania

- Allentown and Lancaster are centers of Pennsylvania Dutch culture, with distinctive foods and crafts for sale.

- At Valley Forge National Historic Park, George Washington and the American army wintered in 1777–78, starving and poorly supplied.

- Gettysburg National Military Park's 30 miles of curving road lead through memorial statues and interpretation of the battle called the "High

Tide of the Confederacy." At adjoining Eisenhower National Historic Park, self-guided tours the retirement home of the 34th president.

- Pocono Mountain resorts have outdoor sports for both summer and winter seasons, including camping, golf, skiing, rafting, and more.

- Hershey, the town that chocolate built, welcomes children with candy factory tours, the Hershey Museum of American Life, and an amusement park.

- Philadelphia's Independence Hall is where the Declaration of Independence was signed. Historic ships are moored at Penn's Landing riverfront park, which includes Independence Seaport Museum. Professional sports, museums, and cultural offerings abound.

- Pittsburgh, growing from river bottoms up steep hills, has two trams climbing Mount Washington, where tourists join commuters. A symphony, professional sports, the Victorian mansion of industrialist Henry Clay Frick, and the Frick Art and Historical Center all attract locals and visitors.

RHODE ISLAND

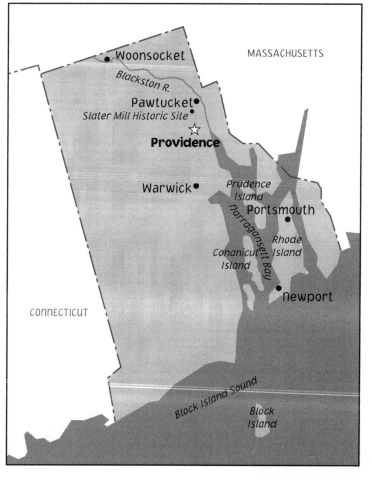

square miles: 1,045 (50th largest)
population: 1,048,319
density: 1,003 people per square mile
capital: Providence
largest city: Providence
statehood: May 29, 1790 (13th state)
nickname: Little Rhody, Ocean State
motto: Hope
bird: Rhode Island red
flower: violet
tree: red maple

Land

North-south Narragansett Bay nearly cuts the smallest state in half, providing most of the state's 40 miles of Atlantic Ocean coast. Thirty-six islands rise in Narragansett Bay, the largest being Rhode Island and, ten miles offshore, Block Island. Coastal lowlands extend from sandy beaches and rocky cliffs to low hills with few trees. The state's northwest third, called western upland, has hills dotted with lakes, ponds, and reservoirs. Jerimoth Hill, 812', is Rhode Island's highest point. The Pawtuxet and Pawtucket rivers shaped Rhode Island's early economy.

Climate

Rhode Island's mild weather does not see extremely high or low temperatures. The average January temperature is 29°F (-2°C), the July average 71°F (22°C). Precipitation is 44" per year, which includes 31" of snow. Hurricanes and tidal waves can occur.

People & History

Settled by religious dissenters from other colonies, Providence Plantations colony settled on the mainland in 1636, and Rhode Island colony at Portsmouth on Rhode Island itself in 1638. Newport was founded in 1639, and Warwick in 1643. Four years later, these colonies formed a loose federation, with capitals at both Providence and Newport until 1900. Today's official name, The State of Rhode Island and Providence Plantations, recalls that history.

Welcoming settlers at first, the Narragansett, Pequot, Wampanoag, and other Algonquin tribes turned against them after their lands were taken. The Narraganseet and Wampanoag tribes joined forces in King Philip's War in 1665, burning almost all the colonists' mainland buildings while survivors fled to Rhode Island. In 1666, British troops ended the war, killing all but 100 warriors.

The colony's official religious tolerance led Quakers and Jews, among religious groups from other colonies and nations, to move there in the 17th and 18th centuries. Touro Synagogue, built in 1763, is the oldest Jewish synagogue in the U.S.

Farmers cleared the land of rocks, building fences with them. The colony prospered in trade, boatbuilding, whaling, and shipping, with Newport rivaling Boston. Using molasses from the West Indies, Rhode Island produced rum for export. After a harmful British tax on molasses in 1764, Rhode Islanders smuggled their product. When a British customs ship, *Gaspee*, chasing smugglers, ran aground off Gaspee Point in 1772, Providence residents burned it. This was the first time American colonists criminally defied Britain.

Although this shipping trade also had included slaves from Africa, Rhode Island, in 1774, became the first colony to ban importing slaves.

Rhode Island also was first colony to declare independence. During the Revolutionary War, British troops occupied Newport; by war's end, Providence had become more important economically.

Rhode Island did not join the United States until the Bill of Rights (see Introduction) was added to the Constitution, and then only when Providence and Newport businessmen won over farmers.

America's first textile mill opened in Pawtucket in 1790, its machinery designed by Englishman Samuel Slater. Rhode Island used its swift rivers to run water-powered factories, and was a leading textile state until mills began moving to southern states (where labor was cheaper) in the 1920s. Costume jewelry-making, begun here in 1794, continues today. In the Victorian era, summer resorts developed on the coast and islands.

In the 19th century, despite religious tolerance, the state refused to let immigrants vote. These included French-Canadian, Italian, Irish, Portuguese, Finnish, and Polish residents. In 1841–42, Thomas Dorr led the Dorr Rebellion, whose participants adopted a state constitution giving all males the right to vote. The effort was crushed, and Dorr—convicted of treason—served one year of a life sentence. Rhode Island then opened voting rights somewhat, but the right to vote in all types of elections was not universal until the mid-20th century.

Industry

Service industries—community and personal, tourism, real estate, finance, and insurance—employ 70% of Rhode Island workers. Jewelry and silverware are the most important manufactured products, followed by fabricated metal items and scientific instruments. Agriculture is a small contributor, mostly from greenhouse and nursery products. Commercial fishing includes fish and shellfish (with lobster the most important catch).

Visting Rhode Island

- Ferries run from Providence to Newport, then on to Block Island. Atop Block Island's 200' Mohegan Bluffs stands Southeast Lighthouse, a Gothic Victorian brick building dating from 1883; it is New England's highest-above-sea-level lighthouse.

- On Rhode Island, Newport's three-mile Cliff Walk National Recreation Trail offers views of Victorian mansions and the ocean; the mansions were summer homes for wealthy 19th-century families. Today ten area estates are museums open to the public.

- Providence is home to The Arcade, the nation's oldest indoor shopping center, dating from 1828. Colonial-era buildings include First Baptist Meeting House (1775), and the home of Declaration signer Stephen Hopkins (built 1743). Rhode Island School of Design's Museum of Art is open to the public, along with Brown University's Wheeler Gallery and List Art Gallery. Rhode Island Philharmonic Orchestra and Trinity Square Repertory Company perform.

SOUTH CAROLINA

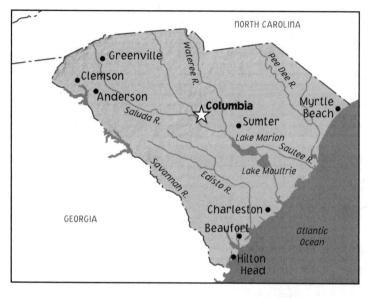

NORTH CAROLINA

- Greenville
- Clemson
- Anderson

Wateree R.

Pee Dee R.

★ **Columbia**

- Sumter

Myrtle Beach

Saluda R.

Lake Marion

Sautee R.

Lake Moultrie

Savannah R.

Edisto R.

GEORGIA

- Charleston
- Beaufort

Atlantic Ocean

- Hilton Head

square miles: 30,109 (40th largest)
population: 4,012,012
density: 133 people per square mile
capital: Columbia
largest city: Columbia
statehood: May 23, 1788 (8th state)
nickname: The Palmetto State
motto: *Dum Spiro, Spero* (While I Breathe, I Hope), and *Animis Opibusque Parati* (Prepared in Mind and Resources)
bird: Carolina wren
flower: Carolina jessamine
tree: palmetto

Land

Reaching inland from the Atlantic Ocean up to 150 miles, the coastal plain covers two-thirds of the state. Its elevation is 100'-300'. Residents call its eastern section the Low Country. Off the southwestern coast lie the Sea Islands, once prime cotton land. (South Carolina has 187 miles of Atlantic Ocean coastline.) The Up Country begins with the Piedmont, rolling to hilly land cut by rivers and lakes. The small section of rugged Blue Ridge Mountains rises mostly to 1,000'-2,000', the highest peak being Sassafras Mountain at 3,560'. The Savannah River forms the Georgia border.

Climate

Weather is mild, and in the Sea Islands subtropical. The long summers are hot and humid, with an average July temperature around 80° F (27° C); average January temperature is 45°–50° F (7°–11° C). Precipitation is concentrated in the summer, averaging 48" a year except in the Blue Ridge Mountains, where it rises to 70".

People & History

A Spaniard was the first European here in 1521, and five years later a Spanish colony lasted only nine months. Settlement began with Charles Towne in 1670, named for Britain's King Charles II, who granted Carolina colony to favorites. It prospered, producing indigo, rice, and cotton. Colonists, at odds with their Lords Proprietors, threw them out in 1719. Ten years later South Carolina split off from North Carolina and became a crown colony, but residents still resented Colonial rule, wanting to control their own budget. This was a major factor in the American Revolution. In 1775, the royal governor fled. After the British conquered Charleston (formerly Charles Towne) in 1780, more Revolutionary War battles occurred here than in any other future state.

With peace came population growth inland, sturdy but poor frontier people. Native Cherokee, Catawba, and Yamasee Indians were driven out by 1800. By 1808, four-fifths of the population lived in the Up Country, but four-fifths of the wealth belonged to Low Country residents. Settlers were mostly English and Irish, with Dutch, French Protestants, and after 1800, many Germans. Five black slaves had been among the first company of settlers; by 1880 African Americans made up 60% of the population (today less than 30%). In 1786, the Up Country majority moved the capital to Columbia.

When U.S. tariffs harmed South Carolina's trade, citizens wanted to secede as early as the 1840s. Other Southern states kept it in the Union until December 20, 1860, when it became the first state to secede. The Civil War's first shots were fired by Confederates at Fort Sumter in Charleston harbor on April 12, 1861.

The war cost South Carolina one-fourth of its men, although battles did not reach the state until 1865. Columbia was burned along with much farmland. The state remained mostly poor and rural until after World War II, with 65% of its farmers as tenants in the 1930s. African Americans were disenfranchised from 1895 until the 1960s.

Industry

The state's textile industry, dating from the 1820s, employed 40% of workers by the 1970s. Cotton was the main crop until the 1950s. Behind textiles come chemical and apparel production, then paper and paper products. Commercial timberlands cover 60% of the state. Atomic energy has been significant, with a large plant on the Savannah River.

Visiting South Carolina

- Myrtle Beach centers the Grand Strand, 60 miles of white sand beach, offering outdoor recreation (golf, deep-sea and surf fishing, snorkeling, diving), amusement parks, and performances (especially country music).

- Charleston has so many historic homes and districts that it offers two self-guided walking tours to such sites as Cabbage Row (model for Porgy and Bess's Catfish Row) and Hibernian Hall (where Irish immigrants gathered beginning in 1799); a Gothic-styled Huguenot (French Protestant) Church, Federal, Adam, and Victorian homes; and boat tours to Fort Sumter National Monument on its man-made island.

- Lexington County Museum Complex in Lexington shows life in the 1830s through buildings, tools, quilts, and demonstrations.

- Brookgreen Gardens near Murrells Inlet features 450 sculptures from the 19th- and 20th centuries amidst native and imported plants and trees.

- World of Energy at Clemson tells the history of power from water to atoms, and includes an aquarium, computer games, and a nature trail.

- Middleton Place and Magnolia Plantation and Gardens, both northwest of Charleston, have homes and gardens showing life before and after the Civil War.

- Walnut Grove Plantation at Roebuck takes visitors back to the 1700s, with main house and outlying buildings.

- Beaufort, the state's second oldest town, preserves many aspects of Atlantic port cities of Colonial and antebellum days.

- Parris Island Museum covers U.S. Marine Corps history.

SOUTH DAKOTA

square miles: 75,885 (16th largest)
population: 754,844
density: 10 people per square mile
capital: Pierre
largest city: Sioux Falls
statehood: November 2, 1889 (40th state)
nickname: The Mount Rushmore State
motto: Under God, People Rule
bird: ring necked pheasant
flower: pasqueflower
tree: Black Hills spruce

Land

Located in North America's center, South Dakota is divided in half by the Missouri River flowing south. The glaciated eastern half holds rich prairies, and the unglaciated western half has hills, buttes, and fertile soil. The Black Hills of the southwest include Harney Peak, the state's highest point at 7,242'. In the northeast, Big Stone Lake, at 962', is the lowest point.

Climate

South Dakota's open countryside experiences hot summers—July temperatures average 74°F (23°C)—and cold winters, with January's average temperature 16°F (-9°C). Annual precipitation averages 18".

People & History

Mound Builders lived along the Big Sioux River 25,000 years ago. By the 1500s, Arikara Indians built fortress villages along the Missouri River. The La Vérendrye brothers were the first Europeans to visit, claiming the land for France in 1742–43. Twenty years later, Spain controlled the land, which it returned to France shortly before that nation sold it to the U.S. in the Louisiana Purchase, 1803.

Sioux Indians, pushed westward by white settlers, took the land from the Arikaras, who also were devasted by European diseases spread by fur traders and trappers. Fort Pierre was built as a fur post in 1817, and the first permanent settlements were Yankton and Vermillion in 1859. Towns developed mostly along the Missouri River until the railroad arrived in the 1870s.

After gold was discovered in the Black Hills in 1874, wild frontier towns such as Deadwood and Lead sprang up. But homesteading brought in more people, mostly northern Europeans. Laws passed in 1917–21 allowed state government to enter insurance, finance, coal mining, and cement making. All but the cement plant failed, leaving the state badly in debt in the 1920s when agricultural prices fell nationwide. The Dust Bowl years of the mid-1930s, along with the Great Depression, left South Dakota unstable until after World War II. Since then, tourism and industry have been encouraged to diversify the economy.

Industry

The Homestake Gold Mine at Lead makes South Dakota the nation's main gold producer. Most agricultural income is from livestock—beef cattle, hogs, and sheep, the state leads all others in raising rye, flax, durum wheat, and sunflower seeds. Meat packing is the major industry, and a small wood-products industry uses ponderosa pine harvested in the Black Hills. Nonelectrical machinery is made along with stone, glass, and clay products.

Visiting South Dakota

• Mount Rushmore National Monument near Rapid City, in the Black Hills, shows the 60' tall heads of Presidents Washington, Jefferson, Theodore Roosevelt, and Lincoln, carved into the mountain by the design and direction of Gutzon Borglum, 1927–41.

• In Rapid City, The Journey Museum offers interactive exhibits on Sioux Indians and Black Hills history. Crazy Horse Memorial is being carved from a mountainside to honor the Oglala Sioux military leader who helped lead the Indian victory at Battle of Little Bighorn. Reptile Gardens show reptiles in natural environments, and Dinosaur Park holds life-size concrete dinosaur models. At Bear Country U.S.A., visitors drive among free-roaming bears.

- In Wind Cave National Park, Wind Cave whistles with the wind: going in during clear weather and out during rainy weather. Bison, pronghorn antelope, elk, and deer graze in the park's prairies.

- Badlands National Park's two units show rugged Black Hills geological formations, wildlife (coyote, buffalo, pronghorn antelope, black-footed ferrets, and deer), and fossil beds.

- Deadwood's Mount Moriah Cemetery includes the graves of frontier characters Calamity Jane and Wild Bill Hickok, who was killed in 1876 while playing poker in a saloon here. Today's visitors can choose among activities such as gold panning, gold mine tours, and 80 casinos.

- In Pierre, the State Capitol is restored to its original appearance. South Dakota Cultural Heritage Center shows the state's natural and human history. Interactive exhibits are featured at the Discovery Center and Aquarium.

- Fort Sisseton State Park holds 14 original buildings from this frontier cavalry post established in 1864.

- Mitchell's Corn Palace was built in 1892 with minarets and spires; each year its surface receives a new mural made of corn, grain, and seeds. Museum of American Indian and Pioneer Life includes art exhibits, Indian artifacts, items from homesteads and the frontier military, a schoolhouse, and horse-drawn farm machinery.

- De Smet is the nearby town of author Laura Ingalls Wilder's Little House on the Prairie fictionalized reminiscences.

TENNESSEE

square miles: 41,217 (34th largest)
population: 5,689,283
density: 138 people per square mile
capital: Nashville
largest city: Memphis
statehood: June 1, 1796 (16th state)
nickname: The Volunteer State
motto: Agriculture and Commerce
bird: mockingbird
flower: iris
tree: tulip poplar

Land

Tennessee is divided into equal thirds. Mountainous East Tennessee includes the Unaka Mountains, part of which is called Great Smoky Mountains. Clingmans Dome, the state's highest point at 6,643', is among them. Middle Tennessee is rolling hills and West Tennessee is an undulating plain dropping to 182' at the Mississippi River. The Tennessee and Cumberland rivers are important for supplying electricity from dams that also created the Great Lakes of the South.

Climate

Tennessee's weather is moderate, with warm summers and cool winters. Average July temperatures are 79°F (26°C) west and 77°F (25°) east; January 48°F (5°C) west and 37°F (3°C) east. West Tennessee's growing season is 100 days longer than East Tennessee's 160 days. Annual 50" of precipitation falls evenly around the state and the seasons.

People & History

When Hernando de Soto crossed this land in 1541 and discovered the Mississippi River near today's Memphis, Cherokee Indians lived in the east, Shawnees in the middle, and Chickasaws in the west.

White settlers moved into Tennessee from other colonies after France ended its claims in 1763 in favor of Britain. They created small frontier farms and towns. The state's western two-thirds supported slavery, and by 1860 one-fourth of its people were African Americans. During the 19th century, few foreign immigrants moved into Tennessee.

East Tennessee's abolitionist mountaineers favored the Union as the Civil War neared, and Tennessee was the last state to secede (and first to rejoin the Union). As Confederate territory shrunk during the Civil War and Union forces moved in, Tennessee saw more major battles than any other state except neighboring Virginia.

Racial segregation settled in after the war, with parallel but separate white and black societies. Black leaders began working in the 1950s for integration of schools, workplaces, and public facilities.

The state's economy did not rebound following the Civil War, and population grew slowly as residents moved away. The Great Depression of the 1930s hit especially hard here, but one of the largest New Deal projects opened the way to a more prosperous future. The Tennessee Valley Authority (TVA) was created in 1933 to develop states of the Tennessee River valley: Kentucky, Virginia, North Carolina, Georgia, Alabama, and Mississippi. Its 46 dams control floods, supply electricity, and allow river traffic from the river's mouth on the Ohio River to headwaters at Knoxville. Recreational lakes also benefit the region's economy.

Each of the "Three Tennessees" has its own hub. Memphis, on the Mississippi, serves the west; it is known especially for influencing jazz. Nashville, in the middle, is the capital of both Tennessee and country western music. East Tennessee boasts both Knoxville, home of the University of Tennessee, and Chattanooga.

Industry

With TVA electricity, Tennessee became one of the nation's top ten manufacturing states, and manufacturing employs a quarter of its workers.

They produce fabricated metals, rubber and plastics, chemicals, automobiles, electronic equipment, and processed food. Printing and publishing are also important. Half the state's workers are employed in service industries. Small farms make up the agricultural sector, which also includes breeding show horses and the Tennessee Walking Horse, developed for ease in riding. Oak Ridge National Laboratory and the TVA employ scientists and engineers seeking advances in electronics and technology.

Visiting Tennessee

- Music lovers of various tastes can wail on revitalized Beale Street in Memphis, "The Home of the Blues," or tour Elvis Presley's Graceland mansion and the Sun (recording) Studio, or celebrate country western at Nashville's Grand Ole Opry and the Country Music Hall of Fame and Museum.

- Soldier/President Andrew Jackson's estate, The Hermitage, is open to visitors near Nashville.

- Half of Great Smoky Mountains National Park crosses the state's eastern border. Nearby Dollywood theme park at Pigeon Forge, owned by country singer Dolly Parton, features craftspeople at work along with rides and other attractions.

- Civil War sites include Fort Donelson National Battlefield, near Dover, the site of the Union's first victory of the war; Shiloh National Military Park, southwest of Savannah, site of a bloody 1862 battle for a railroad; Chattanooga and Chickamauga National Military Park (mostly in Georgia) which commemorate major battles of 1863.

- Knoxville's Frank H. McClung Museum at the University of Tennessee features exhibits ranging from ancient Egypt and early America to decorative arts and medical history; Governor William Blount Mansion, home of a Constitution signer and the state's first governor, is furnished with period pieces from 1792; Knoxville Zoological Gardens exhibits 900 animals.

- Chattanooga is home to the Tennessee Aquarium, displaying ecosystems of the Tennessee and five major world rivers; Chattanooga Choo-Choo has an engine and car from the original 1880 railroad, a restored terminal, and a model railroad; Hunter Museum of American Art, in a 1904 mansion, exhibits art from the 19th century to today.

TEXAS

square miles: 261,797 (2nd largest)
population: 20,851,820
density: 80 people per square mile
capital: Austin
largest city: Houston
statehood: December 29, 1845 (28th state)
nickname: The Lone Star State
motto: Friendship
bird: mockingbird
flower: bluebonnet
tree: pecan

Land

Texas is the largest of the lower 48 states, extending from 367 miles of Gulf of Mexico coastline through fertile coastal plains, then extensions of the Great Plains, to dry High Plains. Guadalupe Peak, in West Texas, is the highest point at 8,751'. The Rio Grande forms Texas's southwestern border.

Climate

Statewide, the January average temperature is 46°F (8°C), and July averages 83°F (28°C). East Texas receives 110" of rainfall per year, West Texas only 44" with recurring droughts, but statewide average precipitation is 27" a year. Around 100 tornadoes occur annually.

People & History

Few native people lived here when the first Spanish explorers appeared in 1528. Military expeditions moved from Mexico into Texas, founding San Antonio by 1718.

In 1821, Stephen Austin founded a colony on a Mexican land grant. Other such settlements followed, and residents were displeased with the restrictive government far away in Mexico City. When Austin traveled there in 1833 seeking statehood, he was imprisoned; by the time he was freed in 1835, Anglo and Mexican Texans were fighting with Mexican troops. The following year Texas declared itself the independent Republic of Texas.

Troops arriving in San Antonio to quash the rebellion surprised Texans who held a Spanish mission, the Alamo. After thirteen days of siege, Mexican soldiers attacked on March 6, 1836, killing all 187 Texans. On April 21, Texas defeated Mexico at San Jacinto.

The Mexican army and occasional Indian raids, plus financial problems, made the residents' lives difficult. In 1845, the Texas and U.S. congresses approved an annexation treaty, with the provision—unique among states—that Texas owned all public lands.

Texas seceded from the union, 1861, and joined the Confederacy during the Civil War. The war's last battle was fought here more than a month after war's end because surrender news had not arrived. Reconstruction lasted until 1874 and saw the rise of the Ku Klux Klan and racial violence. Segregated schools continued until the 1960s.

From the 1860s until railroads arrived in the 1880s, Texas cattle ranchers drove herds to Kansas and Missouri rail centers, a brief cowboy era. Texas's isolation during the Civil War had encouraged building basic factories. Then, in 1901, oil was discovered, and soon dominated the economy. In the 1980s, a large decline in oil and gas prices hurt Texas, which began to recover—and diversify industry—in a few years.

Industry

Texas ranks first among states in railroad and highway mileage, and number of airports. Houston is the busiest of 13 ocean ports on the Gulf of Mexico. A leading agricultural state, Texas is the primary raiser of beef cattle and cotton. It produces a quarter of the nation's oil and a third of its natural gas. Tourism is nearly as great as in Florida and California. Manufactured goods include machinery, chemicals, and food products.

Visiting Texas

- Gulf of Mexico beach resorts include Galveston and Corpus Christi. South Padre Island, offshore near Brownsville, has both resorts and undeveloped coastline.

- Big Bend National Park, on the Rio Grande, offers deserts to explore and the river to raft.

- The Alamo is in downtown San Antonio and is also home to San Antonio Missions National Park, which dates from 1720; River Walk offers boat cruises, shopping, dining, and entertainment, Mexican food and performances. The city has Spanish Colonial architecture, theme parks Fiesta Texas and Seaworld of Texas, plus the Hertzberg Circus Museum, and La Villita Spanish colonial village.

- Fort Worth, once a Chisholm Trail cow town, includes Stockyard Historical District, Amon Carter Museum of Western Art, an excellent zoo, Modern Art Museum, and Forth Worth Museum of Science and History.

- Dallas's Old City Park exhibits restored log cabins and antebellum mansions, there is an active arts community, and the Sixth Floor Museum tells of President Kennedy's life and assassination here.

- Galveston's downtown is Victorian. It has a seaport, long miles of beach, the restored sailing ship *Elissa*, a railroad museum, and historic mansions.

- In Houston, Johnson Space Center exhibits past spacecraft, shows astronauts training, and offers demonstrations. Cowboy skills are demonstrated at George Ranch. Sporting events fill the Astrodome.

- El Paso, at Texas's extreme west and bordering Mexico, is a showplace of the state's blend of Spanish, Indian, and American cultures.

- Austin was founded as Waterloo but renamed to honor Stephen Austin. Live music is the heart of the capital's night life, and visitors can take a paddlewheel boat on Town Lake or tour the pink granite state capitol.

UTAH

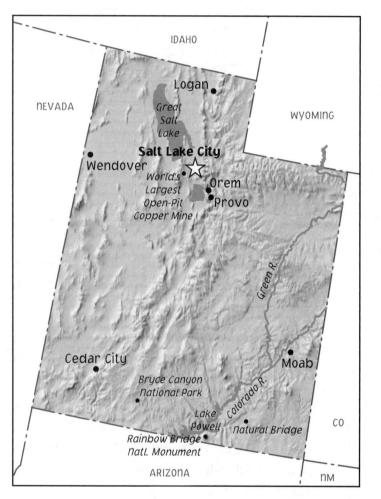

square miles: 82,144 (11th largest)
population: 2,233,169
density: 27 people per square mile
capital: Salt Lake City
largest city: Salt Lake City
statehood: January 4, 1896 (45th state)
nickname: The Beehive State
motto: Industry
bird: seagull
flower: sego lily
tree: blue spruce

Land

Eastern Utah is Colorado Plateau land, and western Utah is part of the desertlike Great Basin, dropping to 2,000' elevation. Through the center run the Rocky Mountains, in which Kings Peak rises in the Uinta range, Utah's highest point at 13,528'. The Great Salt Lake in northwest Utah is a tiny remnant—at about 1,700 square miles—of ancient Lake Bonneville, and is the largest natural lake west of the Mississippi.

Climate

Average January temperature in Utah is 25°F (-4°C), with July averaging 73°F (23°C). Precipitation averages 12" a year statewide, ranging from 5" in the desert to 50" in the mountains. Southwestern Utah has a warmer climate, where cotton once was raised, and is nicknamed Utah's Dixie.

People

Anasazi farming people built cliff dwellings here between 400 and 1250 A.D. , when they moved away. Spanish missionaries in 1776 met Ute and Southern Paiute Indians, some of them farming with irrigation, some following buffalo herds on horseback.

Mexico claimed the land, but sent occasional trading parties rather than settlers. American mountain men arrived in 1811; more came in the 1820s and 1830s at the fur trade's peak. U.S. government parties explored in 1843 and 1845. After Mexico's defeat in the Mexican War, 1848, the U.S. gained control.

White settlement began with emigrants belonging to the Church of Jesus Christ of Latter-day Saints, called Mormons. Founded in New York in 1830, the church then allowed polygamy, which had led to persecution when members tried to settle in Ohio, Illinois, and Missouri. They moved here in 1847, founding Salt Lake City; two years later, they created the State of Deseret (see Nevada) with church leader Brigham Young as governor.

Territorial status was granted as part of the Compromise of 1850 (see Introduction). As Mormon farmers irrigated the desert and took more land, the once-welcoming Ute Indians fought back, in 1853 and again in 1865, when they were defeated in the Black Hawk War.

Federal troops occupied Utah from 1858 until 1861, and again in 1863, to support the non-Mormon governor appointed by President James Buchanan. After the Mormons rejected polygamy and wrote a constitution banning church control of government, Utah was admitted to the union.

In 1869, the first transcontinental railroad had been completed at Promontory Point. With rail shipping, beef and sheep became major exports. Copper mining and smelting were important by the early 1900s. Mining and agriculture were greatly harmed by the 1930s Depression, giving Utah one of the highest rates of unemployment.

Almost all of Utah's people today are of northern European heritage, and 70% are Mormons.

Industry

Two-thirds of Utah is owned by the federal government, including military installations, such as Dugway Proving Ground; Indian reservations;

wilderness areas; and wildlife refuges. Service industries make up the largest sector of the state's economy: tourism, medicine, finance, wholesale and retail trade. Manufacturing produces rocket propulsion systems, automobile airbags, computers, processed foods, metal ores, and fabricated metals. Utah is one of the main sheep-raising states; beef and turkeys are also raised.

Visiting Utah

- Salt Lake City owes its wide streets (excellent for modern traffic) to Brigham Young's requirement that ox-drawn wagons be able to make U-turns. In Temple Square, visitors can take historical tours and hear concerts in the Mormon Tabernacle, but the Mormon Temple is open only to church members. Genealogists can use the church's huge research library. Trolley Square, a shopping and restaurant center, is a state park. The Utah Jazz plays professional basketball.

- At Provo, Alpine Scenic Loop's driving tour reveals views from mountain slopes to Utah Lake. Bridal Veil Falls in Provo Canyon can be viewed from the ground or from an aerial tram.

- Golden Spike National Historic Site, near Promontory, marks where crews building the nation's first transcontinental railroad from the west met crews building it from the east. A museum houses railroad artifacts.

- Near Wendover, Bonneville Salt Flats is the scene of record-setting automobile speeds on the hard natural surface. Bonneville Speedway Museum tells the story.

- Bryce Canyon and Zion national parks in the southwest are filled with amazing rock formations and it also offer scenic drives and hiking.

- Dinosaur National Monument, extending into Colorado, exhibits more than 2,000 dinosaur bones and Indian rock art, and offers guided whitewater rafting.

- Arches National Park near Moab has 2,015 natural stone arches and a 40-mile road wandering among them.

- Canyonlands in the southeast, Utah's largest national park, allows backpacking, mountain biking, horseback riding, river rafting, and boating.

- Ski resorts, including Alta, Park City, and Snowbird, combine excellent snow with low-key atmosphere.

VERMONT

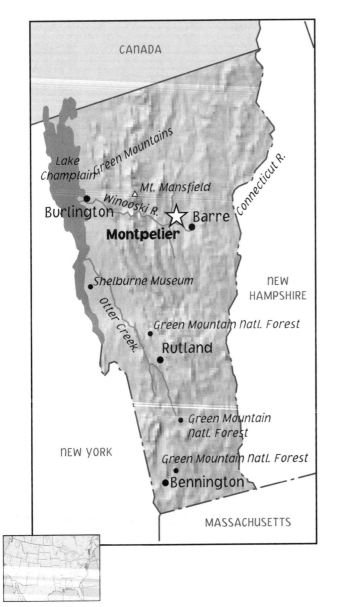

CANADA

Lake Champlain

Green Mountains

Mt. Mansfield

Winooski R.

Burlington

Montpelier

Barre

Connecticut R.

Shelburne Museum

NEW HAMPSHIRE

Otter Creek

Green Mountain Natl. Forest

Rutland

Green Mountain Natl. Forest

NEW YORK

Green Mountain Natl. Forest

Bennington

MASSACHUSETTS

square miles: 9,250 (43rd largest)
population: 608,827
density: 66 people per square mile
capital: Montpelier
largest city: Burlington
statehood: March 4, 1791 (14th state)
nickname: The Green Mountain State
motto: Freedom and Unity
bird: hermit thrush
flower: red clover
tree: sugar maple

Land

The Green Mountains, northernmost U.S. part of the Appalachians, cover central Vermont in a 20- to 36-mile band. On the Massachusetts border is the northern end of the Hoosac Range, whereas the Taconic Range is in southwestern Vermont. Mount Mansfield is the state's highest point, rising to 4,393', and eighty peaks are above 3,000'. Lake Champlain, on the western border, is the state's lowest point at 95' elevation, and its waters drain north into the St. Lawrence River. The Connecticut River forms the eastern New Hampshire border.

147

Climate

Vermont's short summers feature few hot days, usually end in cool nights, and the average July temperature is 68°F (20°C). January's average temperature is 17°F (-8°C). The annual 39" of precipitation includes 80" to 120" of snow.

People & History

Pennacook Confederacy Indians hunted Vermont's woods before the first European, Frenchman Samuel de Champlain, arrived in 1609. French settlers built the state's first white settlement (a temporary one) in 1666 on Isle La Motte, a Lake Champlain island. Nearly eighty years later, Dutch settlers founded Pownal, and English settlers built Fort Dummer at today's Brattleboro.

Vermont was opened to white settlers after the British won the French and Indian War in 1763, and most who moved in came from Connecticut and Massachusetts. Both New York and New Hampshire claimed Vermont, leading to battles between the colonies. When the Continental Congress did not recognize it as a separate state, Vermont declared itself a nation in 1777. It stayed that way until statehood.

The Green Mountain Boys militia formed to fight the "Yorkers" in the decade before the American Revolution. In 1775, that militia turned to the Revolution and captured Britain's Fort Ticonderoga across Lake Champlain, and in 1777 defeated the British in the Battle of Bennington at home. By renewing American hopes, it is considered an early turning point in the war.

Vermont was the only state north of Pennsylvania to experience a Civil War battle when Canada-based Confederates attacked St. Albans.

During the Industrial Revolution, Vermont suffered great emigration of its mostly English-heritage residents to factory states. In the days of water-powered factories, Vermont lacked the important natural resource of swift streams to turn machinery. The state settled into rural, agricultural status, and manufacturing did not begin to develop until the early 20th century. Not until the 1960s did the state's population begin to grow.

Today many residences belong to people who commute to New York, Connecticut, and Massachusetts, and many are vacation and retirement homes.

Industry

Today Vermont depends on manufacturing and tourism, with agriculture comprising only 5% of the state's economy. Milk, maple syrup, Christmas trees, and horses are its main farm products. Most factories have fewer than 50 employees; they produce electronic equipment, machine tools, printing, and wood and paper products (three-fourths of Vermont is hardwood and softwood forest, much of that owned by timber companies). Marble, granite, slate, limestone, and talc are mined. Tourists come to ski and observe skiing competitions, watch maple sugar production as winter turns to spring, enjoy outdoor activities in summer, and view magnificent leaf colors in autumn.

Visiting Vermont

- Shelburne Museum at Shelburne is a reconstructed early American village with 30 historical buildings furnished in period style.

- Smuggler's Notch between Mountain Mansfield and the Sterling Mountains near Stowe gained its romantic name during the War of 1812. When British ships blockaded the ocean, traders brought goods from Boston through the notch.

- Bennington, site of the 1777 colonial victory in the American Revolution, memorializes the battle with a 306' granite tower.

- Proctor holds the world's largest collection of marble and tells the history of Vermont marble quarrying.

- President Calvin Coolidge's birthplace and family home, and an old general store, are in the Plymouth Notch Historic District. Coolidge's plain grave is in the local cemetery.

- Green Mountain National Forest is joined by 34 state forests and 45 state parks. Hiking, fishing, camping, and boating are available.

- Windsor's Old Constitution House, built as a tavern in 1772, is where the state's first constitution was written.

- Barre is the granite city, with the world's largest granite quarries. Visitors can watch as the stone is worked in the world's largest stone-finishing plant. Area cemeteries include intricate and interesting monuments that earlier granite workers carved for themselves.

VIRGINIA

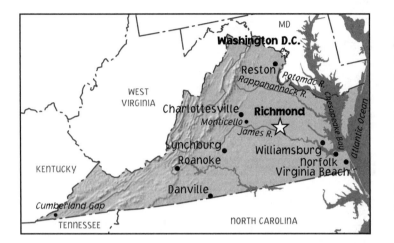

square miles: 39,594 (36th largest)
population: 7,078,515
density: 179 people per square mile
capital: Richmond
largest city: Virginia Beach
statehood date: June 25, 1788 (10th state)
nickname: Old Dominion
motto: Sic Semper Tyrannis (Thus Always to Tyrants)
bird: cardinal
flower: flowering dogwood
tree: flowering dogwood

Land

Virginia shares the southern part of the Delmarva Peninsula between the Chesapeake Bay and the Atlantic Ocean with Delaware and Maryland, and it has 112 miles of bay and ocean coastline. West of the low Tidewater, or coastal plain, Virginia rises into the Piedmont, Blue Ridge Mountains, and Cumberland Plateau. At 5,729', Mount Rogers in the Blue Ridge is the state's highest point. In northwestern Virginia, the Great Valley, drained by the Shenandoah River, has rich soils. The Potomac River forms Virginia's northeast border.

Climate

Tidewater Virginia has warmer summers than inland with its July average temperature 80°F (27°C) while the southwestern mountains average 70°F (21°C). In January, the statewide average is 39°F (4°C). Statewide average precipitation is 40", but the northwestern mountains average 30" while the southeast receives 55".

People & History

A short-lived Spanish settlement was started in future Virginia around 1570, when Powhatan, Susquehanna, and Cherokee Indians lived here.

English colonists founded Jamestown in 1607, beginning tobacco plantations in the Tidewater. Twelve years later, they created the House of Burgesses, the first representative legislature in the Western Hemisphere. Welsh and French Protestant settlers followed, and then Scotch-Irish and German emigrants moved into the Shenandoah Valley.

West of the Tidewater, people created small farms rather than large plantations. Today small farms predominate in the western portion, with more than half Virginia's population concentrated in the extreme east.

In 1763, England banned further settlements in the west, land that France claimed. Virginians, eager to set their own rules, began to support independence, and many became leaders of the American Revolution.

Although the Piedmont and Tidewater depended upon slave labor, Virginia banned the trading of slaves in 1778. It remained a slave state until the Civil War.

Soon after Virginia seceded from the union in 1861, Richmond became the Confederacy's capital. With Richmond as a target for Union troops, and with the Union's capital on its eastern border, Virginia suffered many major Civil War battles. The war ended here when the Confederacy formally surrendered at Appomattox Courthouse in central Virginia, in April 1865.

The war's extensive land damage severely harmed Virginia's agricultural economy, but not until after World War I did the state truly industrialize. World War II and the postwar years brought many military sites that further helped Virginia's economy.

Industry

Government employment is a large part of Virginia's economy, with many federal offices in the northeast across the Potomac from Washington, D.C., and major military bases. Two-thirds of the state is forested, and the wood goes mainly into furniture making. Manufacturing also includes tobacco products, chemicals, textiles, shoes, and electrical and nonelectrical machinery. Tourism to beach resorts and the many historic sites contributes to the economy significantly. The Port of Hampton

Roads (Norfolk, Newport News, and Portsmouth) is a major seaport, and Dulles and Reagan National airports serve Washington.

Visiting Virginia

- Blue Ridge Parkway travels atop mountains for 469 miles between Shenandoah and Great Smoky Mountains national parks. Many visitor centers and exhibits are along the way. Shenandoah Park is a natural area mostly above 2,000' in elevation.

- In Richmond, the capitol was designed by Thomas Jefferson. Monument Avenue displays bronze statues of Confederate leaders and tennis great Arthur Ashe. The Edgar Allan Poe Museum honors the childhood resident and includes the city's oldest home, which dates from 1737. Beside the Museum of the Confederacy stands the White House, Jefferson Davis's home while Confederate president.

- Civil War sites include Richmond, Petersburg, and Manassas (Bull Run) national battlefields, and Fredericksburg and Spotsylvania county battlefields.

- Monticello, Thomas Jefferson's self-designed home in Charlottesville, includes many of the scientist-president's inventions.

- Mount Vernon in Fairfax County has been restored to show the self-supporting farm it was when George and Martha Washington lived there, with the main house exhibiting many of their personal items; the outbuildings include slave quarters.

- Jamestown, Williamsburg, and Yorktown form Virginia's "Historic Triangle." Jamestown Settlement includes living history amid reproductions of the first fort, a Powhatan Indian village, and replicas of the three ships that brought the first colonists. Williamsburg, second Colonial and first state capital, has 80 original and 400 reconstructed buildings from the 1700s; in the Historic Area, costumed workers demonstrate period crafts. Nearby is Busch Gardens amusement park. At Yorktown Battlefield, the British surrendered in 1781 following the last important Revolutionary War battle.

WASHINGTON

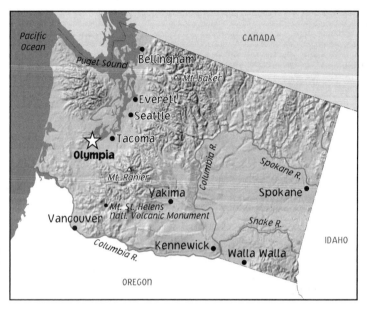

square miles: 66,544 (20th largest)
population: 5,894,121
density: 89 people per square mile
capital: Olympia
largest city: Seattle
statehood: November 11, 1889 (42nd state)
nickname: The Evergreen State
motto: *Alki* (By and By)
bird: willow goldfinch
flower: cosat rhododendron
tree: Western hemlock

Land

On the Pacific Ocean, Washington's 157 miles of general coastline has deep harbors. Puget Sound extends east then south, with ice-free harbors. The Olympic Peninsula in the northwest holds the Olympic Mountains, and south to Oregon extend the Willapa Hills. Inland, the Puget Sound Lowland holds most major cities. To its east, the Cascade Mountains include 14,410' Mount Rainier, the state's highest. The Columbia Basin, cut by the Columbia and Snake rivers, fills central Washington. Northeast, Okanogan Highlands are part of the Rocky Mountains, and the southeast's Blue Mountain region is open, agricultural land.

Climate

January temperatures average 30°F (-1°C), and July averages 66°F (19°C). Precipitation averages 38" statewide, but ranges from 140" on the coast to 16" in the northeast.

People & History

Before Europeans arrived, western Washington was home to Chinook, Makah, Puyallup, Nisqualli, and other Indian nations who lived on seafood and used abundant softwoods to make everyday utensils. In the open east, Spokan, Nez Perce, Okanogan, Yakima, and others followed the buffalo in mobile tepee villages.

Spanish explorers landed in future Washington in 1775 to prevent Russian expansion there. Three years later, Britain became interested in the fur trade, and in 1792, so did Americans. In 1805, the Lewis and Clark Expedition explored along the Columbia River, but British traders were more active until the American Fur Company arrived in 1811.

Britain claimed the region to the Columbia (Washington's southern border), but in 1846 agreed to the 49th parallel—today's Washington/ Canada boundary.

The main Puget Sound settlements began 1845–52. While the lush Willamette Valley attracted wagon trains in the 1850s, Indian resistance in the east prevented Columbia Basin settlement until the late 1850s.

Timbering and salmon canning plus coal and gold discoveries drew more people as railroads arrived (1880s), transporting products to national markets. But just after statehood, an 1890s national economic depression slowed growth.

Washington prospered in the early 20th century. The Great Depression brought large hydroelectric dam projects to the Columbia River. In the 1940s, aircraft production rose to prominence, adding spacecraft in the 1960s. Downturns in military spending harmed the whole state, but high-tech industries in the 1980s diversified the economy.

Industry

Manufacturing is one-fifth of Washington's economy, and the largest employer is an airplane/spacecraft maker. In a related industry, imported bauxite is processed into aluminum at seven smelters. Computers and software are significant products, and three major shipbuilding yards exist. Forest industries utilize Western hemlock and Douglas fir. Agriculture is a small contributor, but Washington leads the states in apple growing. Commercial ocean fishing catches salmon, halibut, and shellfish. Outdoor-recreation tourism has grown greatly in recent decades.

Visiting Washington

- Seattle can be viewed from the top of the Space Needle; Pike Place Market has varied food and other shops; Pioneer Square Historic District's cobblestoned streets run amid outdoor cafes; Klondike Gold Rush National Historical Park recalls Seattle's function as supply point for 1890s prospectors; museums include the Seattle Art Museum, Nordic Heritage Museum, and Museum of History and Industry.

- Mount Rainier National Park surrounds Mount Rainier, which is visible from Seattle on a clear day; glaciers top the mountain, wildlife wander lower forests, and fall colors are a special draw. Trails, downhill skiing, and a scenic railroad trip are available.

- Olympic National Park and the Olympic Peninsula hold the world's only coniferous rain forest (nourished by 130" of annual rain), and 50-plus miles of beach. Visitors can river or sea kayak, raft, hike, fish, surf, backpack, and fly over. Makah Cultural Museum and Research Center exhibits a 16th-century Makah Indian village. Wildlife includes Roosevelt elk, black bears, cougars—amid 25-story-tall Douglas firs.

- Mount St. Helens National Volcanic Monument (visitor center north of Longview), covers the volcano's 1980 explosion and mudslides, which took 1,300' off the mountain's then-cone-shaped top, leaving it 8,365' high. Helicopter rides, nature hikes, fishing, and camping are available.

- Spokane in eastern Washington has a hand-carved 1909 carousel in Riverfront Park plus a monorail ride, contemporary outdoor statuary, and concerts in an amphitheater. Area wineries are open to tours. Manito Park includes a Japanese formal garden and rose garden.

- Whale-watching boats leave the San Juan Islands in Puget Sound, where people bike, sea kayak, bird-watch, scuba dive, golf, and fish.

- Lake Roosevelt National Recreation Area behind Grand Coulee Dam has a 151-mile-long lake with sandy beaches, and offers dam tours, windsurfing, camping, canoeing, hiking, and fishing.

WEST VIRGINIA

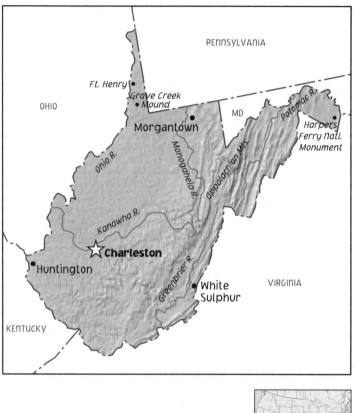

square miles: 24,078 (41st largest)
population: 1,808,344
density: 75 people per square mile
capital: Charleston
largest city: Charleston
statehood: June 20, 1863 (35th state)
nickname: Mountain State
motto: *Montani Semper Liberi* (Mountaineers Are Always Free)
bird: cardinal
flower: rhododendron
tree: sugar maple

Land

All of West Virginia lies within the Appalachian Mountains, including (east to west) the Blue Ridge Mountains, the Ridge and Valley region, the Allegheny Mountains, and the Appalachian Plateau that covers three-fifths of the state. In the Alleghenies, Spruce Knob at 4,863' is the state's highest point.

Climate

Precipitation is West Virginia's notable climate feature, with 48" spread evenly through the year, and more at higher elevations. January's temperatures average 32°F (0°C), and July's 72°F (22°C). The state's steep mountainsides often cause flash floods.

People & History

Stone Age hunters chased woolly mammoths here 14,000 years ago. Beginning about 3,000 years ago, Adena mound builders settled the area for ten centuries. Their earthworks still exist. By the mid-1600s, this land was hunting and battle ground without Indian residents.

White settlers reached the Blue Ridge in this western part of Virginia Colony in 1670. Their numbers increased gradually because of difficulties that included France and England's conflict over ownership, Indian warfare, and the rugged mountains. The hilly land was suited to small farms rather than large plantations, and most people who moved here opposed slavery, so there few slaves.

The very different lifestyle and concerns of western citizens caused a movement toward separate statehood as early as 1776. When Virginians met in 1861 to decide whether to secede from the Union, delegates from the west voted no. Western Virginia lived under its separate Government of Virginia until West Virginia became a state two years later.

With the arrival of railroads in the 1870s, West Virginia drew on its natural resources to supply northern factories with coal, timber, oil, gas, and salt. Efforts to form labor unions in the coal mines led to violence from 1912 to 1921, with both the U.S. Army and the National Guard called in a total of six times.

Industry

Coal from underground mines has been West Virginia's most important natural resource for more than a century, with major reserves still remaining; oil is also produced. The state ranks third in production of soft coal nationally, and all mining accounts for 10% of the state product. Transport of coal by truck, river barge, and railway is another important economic activity. Electricity is produced for local use and export to other states. Manufacturing includes chemicals, petrochemicals, metals (including steel) and fabricated metal products, glass, pottery from local clays, and wood products. Salt has been produced here since the earliest settlements. Hunting and fishing attract visitors, and the state has the East's largest federally protected river system, which also attracts recreationists.

Visiting West Virginia

- Two thousand miles of swift Appalachian Mountain streams, along with slower waters, serve whitewater rafters from beginners to the highly experienced.

- Harpers Ferry National Historical Park holds the Federal Arsenal and Armory that was built in 1796, which then supplied the Lewis and Clark Expedition seven years later. In 1859, abolitionist John Brown and his followers briefly captured the arsenal, intending to arm slaves so they could rebel against their masters.

- Beckley's Exhibition Coal Mine takes visitors in converted coal cars with retired miners describing their work. Outdoor summer pageants are Honey in the Rock (about West Virginia's creation) and Hatfields and McCoys (about a long-time mountain-family feud).

- Following an old logging railroad, Cass Scenic Railroad's steam locomotives take passengers to Bald Knob.

- Blennerhassett Island Historical State Park in the Ohio River downstream from Parkersburg can be reached by paddlewheeler now. In 1806, owner Herman Blennerhassett and former U.S. Vice President Aaron Burr supposedly plotted here to make the American Southwest a separate nation, and possibly invade Mexico. Blennerhasset Mansion is reconstructed.

- Since people have lived here, they have enjoyed the waters at White Sulphur Springs, and today the Greenbrier Resort also offers a challenging golf course.

- Wheeling on the Ohio River grew from Fort Henry, built 1769–70. It was the scene of the last American Revolution battle—after war had ended but before word reached here. The 1849 Suspension Bridge was the world's longest when built. Restored Centre Market, Victorian homes, Wheeling Downs Greyhound Racetrack, and nearby Oglebay Resort are Wheeling's varied attractions.

- In Charleston, the state capitol's 300' dome shines with gold leaf. Once home to Daniel Boone, the town today offers paddlewheel river cruises, a greyhound racetrack, the state museum, an art museum and a planetarium.

WISCONSIN

square miles: 54,310 (26th largest)
population: 5,363,675
density: 99 people per square mile
capital: Madison
largest city: Milwaukee
statehood: May 29, 1848 (30th state)
nickname: The Badger State
motto: Forward
bird: robin
flower: wood violet
tree: sugar maple

Land

Wisconsin has 381 miles of Lake Michigan shoreline (where the state's lowest point, 581', lies), and 292 miles on Lake Superior. From the Superior lowland in the north, the land rises to the northern highland, which includes Timms Hill, 1,952', Wisconsin's highest point, and covers the state's northern third. Farther south is the central sandy plain. The eastern ridge and lowland region has few hills. All the land except the southwest was scraped by glaciers, so its river-cut hills show the state's greatest relief.

Climate

Wisconsin's humid continental climate is moderated in the east by the Great Lakes. Summers are mild to warm with a statewide July average temperature of 70°F (21°C). Cold snowy winters see a January average temperature of 14°F (-10°C). Statewide, annual precipitation averages 31", but the snowier north sees 36" a year.

People

People moved into future Wisconsin as the last glaciers retreated 13,000 years ago. After these Stone Age hunters, Indians who made copper tools lived here 6,000 years ago, followed by Mound-Builder cultures from 100 B.C. to about 1000 A.D. In the 1600s, Winnebago Indians lived in villages from which they hunted and fished; other nations in the state included the Sioux, Chippewa, Miami, Potowatami, and Kickapoo.

French fur trader Jean Nicolet reached the Green Bay area in 1634; Wisconsin belonged to France until 1763. Britain controlled it for 110 years until after the American Revolution. Mostly French, English, and Scottish settlers lived here until after the War of 1812. U.S. forts built at Green Bay and Prairie du Chien in 1816 began American settlement. Discovery of lead deposits in the southwest twenty years later brought miners, followed by farmers from eastern states, northern Europe, and Scandinavia.

Farming in the southern two-thirds of the state specialized, with Wisconsin becoming the leading dairy state by 1920.

In the mid-20th century, many African Americans moved to Wisconsin, especially to its cities. The Indian population is centered mainly in Menominee County's Menominee Indian Reservation.

The antislavery Republican Party was founded here in 1854, and the Progressive Party—formed in 1900—put Wisconsin in the forefront of legislation to protect individuals from laws influenced by corporations. This political philosophy continues today.

Industry

Livestock and dairy farms account for 80% of Wisconsin's agricultural income. A major industrial state, Wisconsin manufactures wood and paper products, beer (especially in Milwaukee), nonelectrical machinery, electrical equipment, and transportation equipment. Historic sites, outdoor recreation, and deer-, rabbit-, and squirrel-hunting seasons attract many visitors. Milwaukee, Green Bay, and Superior are important Great Lakes shipping ports, and ports along the Mississippi River serve barge traffic.

Visiting Wisconsin

- Green Bay has restored Fort Howard (1816), and also offers the National Railroad Museum, a botanical garden, and the Green Bay Packers Hall of Fame.

- Children of all ages visit Baraboo, birthplace of Ringling Brothers Circus and home to Circus World Museum, for circus wagon, costume, poster exhibits, and performances. At nearby Devil's Lake State Park, sheer cliffs rise above a spring-fed lake.

- Milwaukee's buildings include the Milwaukee Art Center (designed by Eero Saarinen), Annunciation Greek Orthodox Church (by Frank Lloyd Wright), the restored Pabst Theater, and the 1890s Pabst mansion with a 15th-century chapel moved from France. Museums include Milwaukee Public Museum, with street-scene exhibits, the International Clown Hall of Fame, and Betty Brinn Children's Museum. Breweries present tours; professional baseball and basketball games are available.

- Visitors to Wisconsin Dells, north of Baraboo, can take boat tours on the Wisconsin River through scenic cliffs, choose among water parks, watch a water-thrill show on Lake Delton, visit costumed characters at Storybook Gardens, and explore the Winnebago Indian Museum.

- Hayward displays lumberjack skills, especially during July's Lumberjack World Championships. National Fresh Water Fishing Hall of Fame's

tackle and trophy exhibits are housed in a four-story replica of a muskellunge fish.

- Land O'Lakes is the center for 100 lakes' resorts and fishing.

- Madison, capital and home to the University of Wisconsin, offers the State Historical Museum, Elvehjem Museum of Art, Washburn Observatory, Geology Museum, Arboretum, and many Frank Lloyd Wright buildings.

- At Oshkosh, Experimental Aircraft Association's museum holds the world's largest collection of flying machines; in July the Fly-In Convention includes airshows and competitions. Paine Art Center and Arboretum is in a Tudor mansion. Grand Opera House has been restored to its 1883 appearance.

- Door County includes Lake Michigan beaches, lighthouses, working ports, hiking, quaint towns, water sports, and winter sports.

WYOMING

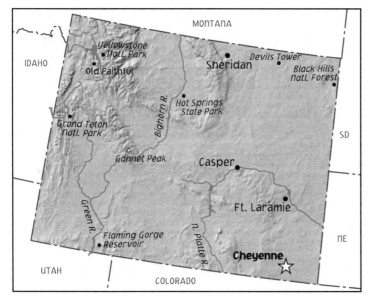

square miles: 97,100 (9th largest)
population: 493,782
density: 5 people per square mile
capital: Cheyenne
largest city: Casper
statehood: July 10, 1890 (44th state)
nickname: The Equality State
motto: Equal Rights
bird: meadowlark
flower: Indian paintbrush
tree: cottonwood

Land

Several ranges of the Rocky Mountains pass through Wyoming, north to south. Gannett Peak in the Wind River range is the state's highest point, at 13,804'. Between the mountains are basins of dry, treeless land. Eastern Wyoming is the end of the Great Plains, and part of the Black Hills reaches into the northeastern corner, where the Belle Fourche River is Wyoming's lowest point at 3,100'.

Climate

With the high mountains wringing moisture from rising clouds, most of Wyoming is semiarid, receiving only 13" of precipitation per year. This dryness makes the state's relatively low temperatures more comfortable—in July the average is 67°F (19°C), and in January 19°F (-7°C).

People & History

Ancient people hunted here at least 11,000 years ago, replaced in the 1700s by Indians who followed buffalo and other game: mostly Shoshone and Arapaho, but also Crow, Cheyenne, Sioux, Kiowa, Bannock, and Blackfeet. No one knows whether the French Vérendrye party, coming from the northeast, or Spaniards from the southwest reached future Wyoming first.

The first confirmed visitor was John Colter, a former Lewis and Clark Expedition member, who stayed in the Rockies in 1806. In American settlements, the following year, he described rich wildlife, boiling springs and mudpots, and geysers. People laughed at "Colter's Hell"; today it is Yellowstone National Park.

Mountain men soon followed, trapping and trading for beaver and buffalo. Their rendezvous, where free-roaming trappers met once a year to sell furs, was held fifteen times in Wyoming, beginning in 1824.

In 1834, their first trading post was built; sixteen years later the army bought it, renaming it Fort Laramie, for soldiers protecting wagon trains on the Oregon Trail through Wyoming. In 1867, the first transcontinental railroad came through Wyoming. Settlements began along its rails, and cattle and sheep ranching soon followed.

Industrial mineral deposits were Wyoming's riches. Coal was found in 1867, then major oil and natural gas deposits in 1912. In 1904, the first dude ranch was founded 32 years after America's first national park, Yellowstone, opened. Recent years have seen less demand for American uranium, oil, and coal, causing Wyoming's population to decline.

Industry

Service industries, including processing and transporting Wyoming's many natural resources, make up the largest economic sector. Wyoming leads the states in soft coal (which underlies 40% of its land), but oil brings in more revenue, along with uranium, bentonite clay (used in chemical manufacturing), soda ash (used to make paper, chemicals, and glass), and some gold. Wyoming makes more of its income from mining than does any other state. Government is an important income source, with the federal government owning one-third of Wyoming. Much of that is set aside for parks, but also includes forests that are logged.

Visiting Wyoming

- Most people visit Yellowstone National Park in summer, but portions are open in winter for cross-country skiing, snowcoach tours, and snowmobiling. Old Faithful Geyser is the most famous of many large and small geysers. Buffalo, elk, and deer frequently are spotted from two-lane roads. Hiking includes short trails, interpreted trails, and back country.

- Grand Teton National Park, bordering Yellowstone to the south, includes the abrupt rise of the Teton range of the Rockies 2,100' above the floor of adjoining Jackson Hole valley. Hiking, climbing, scenic drives, guided horseback rides, and flatwater Snake River rafting are available. Nearby Jackson is a ski resort with tram rides in summer, Wild West entertainment and rodeos.

- Fort Laramie National Historic Site, near Fort Laramie, has ruins, and original and restored buildings from the fur, later military, post that was a Pony Express station in 1860-61.

- Cheyenne celebrates its frontier days of stagecoaches, range wars, rustlers, and gunfighters, but also includes guided tours of the Corinthian-style state capitol, and more of the Old (and New) West story at Wyoming State Museum.

- At Fossil Butte National Monument, 50-million-year-old fossilized freshwater fish were discovered in 1856; today it has driving and hiking tours.

- Jackson is home to a destination ski resort, Teton Village, with summer tram rides. The National Elk Refuge was created to protect animals that migrated to this area each winter. The National Wildlife Art Museum's modern building houses an important collection. Summer entertainment includes mock shootouts, Grant Teton Music Festival's symphony programs, and river rafting.

- Devils Tower National Monument, the nation's first in 1906, holds an 867' high volcanic remnant shaped like a cone with a flat top.

DISTRICT OF COLUMBIA

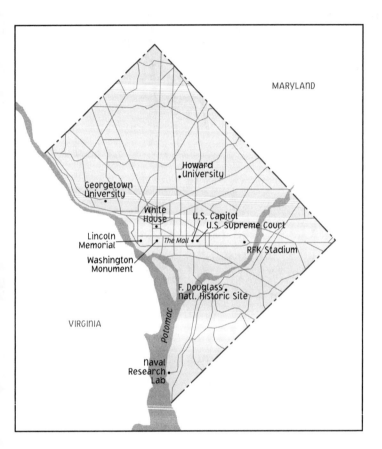

square miles: 68
population: 529,000
density: 7,780 people per square mile

Land

The diamond-shaped area covers the same land as the city of Washington, the nation's capital city, and includes the conjunction of the Potomac and Anacostia rivers. The land was given to the federal government by Maryland (today its border on three sides) and Virginia.

Climate

The District of Columbia's January temperatures average 37°F (3°C); its hot, humid Julys average 78°F (26°C). Annual precipitation averages 50", including occasional snowfalls.

People & History

George Washington (a Virginian) selected this site for the nation's capital. The land was donated so the capital wouldn't "belong" to a state. Congress accepted the site in 1790, and French engineer Pierre Charles L'Enfant completely designed the city of Washington before it was built. Second president John Adams moved the capital from Philadelphia (one of nine temporary capitals) to Washington, D.C. in 1800. The District has a nonvoting representative in Congress. Since 1961, residents have been allowed to vote in presidential elections. African-Americans form the majority of residents, with growing Asian and Hispanic populations.

Visiting Washington, D.C.

- Guided tours take visitors through the main-floor public rooms of the president's residence, the White House, and also the U.S. Capitol.

- The Smithsonian Institution has been called America's attic, and includes 14 themed museums from National Air and Space Museum to the National Zoo and art museums.

- Paddle boats are one way visitors can view the Tidal Basin, cherry trees on its shore, and the Jefferson Memorial.

- The Lincoln Memorial holds a 19' statue of Abraham Lincoln, and inscriptions quoting his writings. Ford's Theater has been restored to its appearance on the night Lincoln was assassinated there, and offers tours as well as theatrical performances.

- On the Mall, there are exhibits for people with many interests at the National Museum of American History, National Gallery of Art, National Museum of African Art, Sackler Gallery of Asian Art, U.S. Holocaust Memorial Museum, National Portrait Gallery, Museum of Women in the Arts, among others.

TERRITORIES OF THE UNITED STATES

Territories are associated with the U.S. but do not have as much self-government as the states. Residents of territories are U.S. nationals free to travel anywhere in the country, but they cannot vote in presidential elections, and do not pay federal income tax.

The "incorporated" version of a territory—a state in the making, living under the Constitution—has the most rights, including electing its own legislature and nonvoting representative to Congress. Before statehood, all but nineteen states were territories; the ones who were not are the thirteen colonies, California, Kentucky, Maine, Texas, Vermont, and West Virginia. At the end of the 20th century, the U.S. had no incorporated territories.

A commonwealth operates under its own constitution, and elects a representative to Congress who can vote at the committee level. Puerto Rico and the Northern Mariana Islands are commonwealths.

Unincorporated territories can be organized or unorganized, and are governed by the executive branch of the U.S. government. Residents have fundamental rights, but not necessarily all those guaranteed under the U.S. Constitution. Organized territories are governed by congressional law that acts as their state constitutions; they have their own legislatures and send nonvoting delegates to Congress. Guam and the Virgin Islands of the United States are organized territories.

American Samoa is an unorganized territory under jurisdiction of the U.S. Department of the Interior, although residents elect their own governor and legislature and send a nonvoting delegate to Congress. The remaining unorganized territories are the Pacific Islands of Baker, Howland, Jarvis, Johnston Atoll, Kingman Reef, Midway, Navassa, Palmyra, and Wake.

Commonwealth of Puerto Rico

square miles: 3,425
population: 3,731,000
density: 1089 people per square mile
capital: San Juan
largest city: San Juan
commonwealth date: July 25, 1952

Puerto Rico lies about 1,000 miles southeast of Florida, bordered on the north by the Atlantic Ocean and on the south by the Caribbean Sea. One of the largest islands between the U.S. and

South America, it has a total coastline of 311 miles. From sea level, it rises to 4,389' on Cerro de Punta.

Sea breezes cool the summers so that the July average temperature of 80°F (27°C) is little higher than January's average 73°F (23°C). Freezing conditions never occur. Rainfall averages 37" a year in the south and 70" in the north. Severe hurricanes occur about once every 10 years.

Christopher Columbus arrived here in 1493, claiming it for Spain. Possession passed from Spain to the United States in 1898, following the Spanish-American War. Spanish is the main language, but English also is an official language.

Service industries—especially tourism—account for half of Puerto Rico's economy, but manufacturing is the single most important industry. The main products from its 1,800 factories are chemicals, pharmaceuticals, food products, electrical equipment. Coffee is the most important crop, followed by sugar cane.

Commonwealth of the Northern Mariana Islands

square miles: 179
population: 64,000
density: 236 people per square mile
capital: Saipan Island
largest city: Garapan
commonwealth date: November 3, 1986

Part of Micronesia, these islands spread over 337 miles in the Pacific Ocean, south of Japan and east of the Philippines. The largest islands are Saipan, where most people live, Tinian, and Rota.

Spain claimed the Marianas (including Guam) in 1565. After defeat in the Spanish-American War, in 1898, Spain turned over Guam to the United States but kept the Northern Marianas, selling them to Germany in 1899. Japan captured them in 1914, during World War I, and ruled them until 1944, when the U.S. seized them. Government is the largest employer, and tourism the biggest sector of the economy.

Territory of Guam

square miles: 209
population: 146,000
density: 698 people per square mile
capital: Agana
largest city: Agana
territory date: August 1, 1950

Guam, southernmost of the Mariana Islands, is 1,300 miles east of the Philippine Islands, bordered on the west by the Philippine Sea and on the east by the Pacific Ocean. A volcanic island, it often experiences earthquakes. Mount Lamlam, at 1,332', is the highest point.

The mean annual temperature is 78°F (26°C), and typhoon season is from June through December.

The native people, Chamorros, arrived here from Asia around 3000 B.C.; today they are about a third of the population. Spain claimed Guam in 1565, instituting government in 1668, and Agana (population 1,139) has buildings dating from early Spanish days. Following its defeat in the Spanish-American War, 1898, Spain turned Guam over to the United States. It was governed by the U.S. Navy from 1917 to 1950.

Japan attacked Guam on December 8, 1941, and captured the island two days later. In 1944, U.S. forces recaptured Guam. Since the war, the U.S. military has occupied one-third of the land, operating navy and air force bases; about a tenth of today's population are U.S. military personnel. Many Guamanians are employed by the military, the second largest component of the territory's economy. Most important is tourism—mainly Japanese visitors.

Virgin Islands of the United States

square miles: 132 (main islands)
population: 118,000
density: 874 people per square mile
capital: Charlotte Amalie
largest city: Charlotte Amalie
commonwealth date: 1954

The Virgin Islands, between the Atlantic Ocean and the Caribbean Sea east of Puerto Rico, are divided between the U.S. and Great Britain. The U.S. Virgin Islands include St. Croix (80 square miles), St. Thomas (32 square miles), St. John (20 square miles), and 65 islets.

Columbus visited and named the islands on his 1493 trip to the New World, claiming them for Spain. No Spanish settlement occurred. Denmark took possession of St. Thomas in the 1670s, St. John in 1717, and bought St. Croix from France in 1733. Denmark sold the Danish West Indies to the U.S. in 1917.

During World War II, the islands were an important base for the U.S. to protect the Panama Canal from invasion by Germany and Italy.

Tourism employs half the three main islands' people, who serve more than 1 million visitors a year. Virgin Islands National Park covers two-thirds of St. John and part of St. Thomas; mules and jeeps provide park transportation, and remains of Danish plantations and sugar refineries can be viewed. Recent years have seen the opening of an aluminum ore and an oil refinery, and factories producing perfume, watches, and thermometers. Most products are exported to the U.S.

American Samoa

square miles: 77 (on seven islands)
population: 60,000
density: 607 people per square mile
capital: Pago Pago
largest city: Pago Pago
territory date: 1899

Polynesian residents were joined by European settlers in 1830, and the islands agreed to be a U.S. Navy coaling station in 1872. They came under U.S. control in 1899 under a treaty among Germany, Britain, and the U.S. American Samoa was governed by the U.S. Navy until 1951, and since then by the Department of the Interior.

Tuna canning is economically important along with government and tourism. American Samoa National Park holds wildlife living in a tropical rainforest and around a coral reef.

Wake Island

Wake Island was named for a British sea captain who landed there in 1796, although Spaniards had seen it in 1568. In the west-central Pacific Ocean, it covers 3 square miles on the coral islands of Wake, Wilkes, and Peale, with about 300 residents, including U.S. weather and oceanographic scientists. It became a territory in 1898, and the U.S. was building an airbase there when World War II began. It was captured and held by the Japanese from 1941 to 1945, and today serves as an emergency airplane stop. It is governed by the U.S. Air Force.

Midway

On the islets of Eastern and Sand, surrounded by a coral atoll, Midway has fewer than 500 residents on 6 square miles. Located about 2,000 miles northwest of Hawaii, it was annexed by the United States in 1867. Beginning in 1935, it served as an airplane refueling base. The Battle of Midway, June 4–6, 1942, between U.S. and Japan, was a turning point in World War II, and one of the war's major naval battles. The U.S. victory prevented Japan from capturing Midway, which it could use to attack Hawaii; in the battle, Japan lost five ships and the U.S. lost two; the islands have been an important naval base since. Today Midway, governed by the U.S. Navy, is a base for air-sea rescue forces.

INDEX

Adams, John, 166
Alabama, 15–17
Alamo, 142, 143
Alaska, 7, 9, 18–20
Alaska Highway, 19
Albany, New York, 108, 109, 110
Albuquerque, New Mexico, 105, 107
aluminum production, 154
American Indians. See Native Americans
American Revolution. See Revolutionary War
American Samoa, 170
Anchorage, Alaska, 18, 20
Annapolis, Maryland, 72, 73, 74
Appalachian Mountains, 8
Appomattox Courthouse, 151
Arizona, 11, 21–23
Arkansas, 24–26
Astor, John Jacob, 124
Atlanta, Georgia, 42, 43, 44
Atlantic Ocean, 7, 8
Augusta, Maine, 69, 71
Austin, Stephen, 142
Austin, Texas, 141, 143
automobile manufacturing, 79
Avery Island, 68

Baltimore, Maryland, 72, 73, 74
Baton Rouge, Louisiana, 66, 67, 68
Battle of Fallen Timbers, 55, 79, 118

Battle of Lake Erie, 79, 118
Battle of Tippecanoe, 55
Bering, Vitus, 19
Billings, Montana, 90, 92
Bill of Rights, 10
Bismarck, North Dakota, 114, 115, 116
Black Hawk War, 52, 58, 145
Boise, Idaho, 48, 50
Boone, Daniel, 64
Boston, Massachusetts, 75, 76, 77
Boston Tea Party, 76
Bridger, Jim, 31
Brulé, Étienne, 79
Buchanan, James, 145
Bunker Hill battle (Revolutionary War), 76
Butte, Montana, 90, 91

Cabeza de Vaca, Álvar Núñez, 40
Cabot, John, 70
Cadillac, Antoine de la Mothe, 79
Cadillac Mountain, 70
California, 11, 12, 27–29, 167
Calvert, Leonard, 73
Cape Cod, Massachusetts, 75, 76, 77
Carson City, Nevada, 96, 98
Cascade Mountains, 8
Champlain, Samuel de, 70, 76, 148
Charleston, West Virginia, 156, 158
Chesapeake Bay, 72, 74

Cheyenne, Wyoming, 162, 164
Chicago, Illinois, 51, 52, 53
Church of Jesus Christ of Latter-day Saints, 145
civil rights movement, 13, 16–17, 85
Civil War, 11–12
  Appomattox Courthouse, 151
  Iowa role, 58
  Kentucky role, 64
  Sharpsburg battle, 73
  Vicksburg, Mississippi, 85
Cleveland, Ohio, 118, 119
coal mining, 64–65, 115, 127, 128, 157, 163
Coast Ranges, 8
Cold War, 13
colonies, original thirteen, 9, 167
Colorado, 30–32
Colt, Samuel, 34
Colter, John, 163
Columbia, South Carolina, 132
Columbia River, 8, 125
Columbus, Christopher, 168, 169
Columbus, Ohio, 117, 119
commonwealth defined, 167
Commonwealth of Puerto Rico, 167–68
Commonwealth of the Northern Mariana
  Islands, 168
Compromise of 1850, 11
Concord, Massachusetts, 76
Concord, New Hampshire, 99
Confederate States of America, 11–12
Connecticut, 9, 33–35
Connecticut Compromise, 34
Constitution, 10, 37

Continental Divide, 8, 10
Cook, James, 46, 124
copper mining, 22, 23, 91
Cornwallis, Charles, 103
Coronado, Francisco Vásquez de, 61, 121
cotton production, 15, 17, 25
Cumberland Gap, 64
Custer, George, 92, 116

dairy farming, 160
Death Valley, 8
Declaration of Independence, 9, 37, 127
Delaware, 9, 36–38
de Monts, Pierre, 70
Denver, Colorado, 30, 32
Des Moines, Iowa, 57, 58, 59
de Soto, Hernando, 16, 25, 43, 67, 85, 139
Detroit, Michigan, 78, 79, 80
District of Columbia, 11, 73, 165–66
Dorr, Thomas, 131
Dover, Delaware, 36
Drake, Francis, 124
Dubuque, Julien, 58
Du Pont, Eleuthére Irenée, 37
Dust Bowl, 13

Economy of the United States, 13–14
Edison, Thomas, 104
Emancipation Proclamation, 12

Fall Line, 8
Florida, 39–41
Ford, Henry, 79

Fort McHenry (Maryland), 73
Frankfort, Kentucky, 63

Gadsden Purchase, 22, 106
geographic center of North America, 60–61
Georgia, 9, 42–44
Geronimo, 106
gold mining, 22, 97, 136
gold rush, 11, 28, 31, 49, 91
Grand Canyon, 8, 23
Gray, Robert, 124
Great Basin, 8
Great Depression, 13
Great Lakes, 8
Great Plains, 8, 12
Green Mountain Boys, 148
Guam, Territory of, 168–69
Gulf of Mexico, 7, 8
gun manufacturing, 34–35
gunpowder manufacturing, 37

Hamilton, Alexander, 103
Harrisburg, Pennsylvania, 126
Hartford, Connecticut, 33, 34, 35
Hawaii, 7, 13, 45–47
Helena, Montana, 90, 92
Hells Canyon, 8, 50
Homestead Act, 12, 115
Honolulu, Hawaii, 45, 47
Hoover Dam, 28, 97, 98
Houston, Texas, 141, 143
Hudson, Henry, 103, 109

Idaho, 48–50
Illinois, 51–53
immigration, 9, 12
incorporated territories defined, 167
Indiana, 54–56
Indianapolis, Indiana, 54, 55, 56
Indian reservations, 12, 23, 43, 112, 121
Indian Territory, 10, 121
Indigenous People. See Native Americans
Industrial Revolution, 76, 100
Iowa, 57–59

Jackson, Mississippi, 84, 85
Jefferson, Thomas, 10, 152
Jefferson City, Missouri, 87, 89
Jim Crow Laws, 16
Jolliet, Louis, 52, 58
Juneau, Alaska, 18, 20

Kamehameha I, King, 46
Kamehameha III, King, 46
Kansas, 60–62
Kansas City, Missouri, 87, 88, 89
Kansas-Nebraska Act, 10, 61
Kentucky, 11, 63–65, 167
Key, Francis Scott, 73
King Philip's War, 76, 100, 130
Kino, Father, 22
Korean War, 13

labor unions, 52
Lake Erie, 118
Lansing, Michigan, 78, 79, 80

La Salle, René-Robert Cavelier, 25, 52, 55, 67
Las Vegas, Nevada, 96, 98
La Vérendrye, Sieur de, 115, 136, 163
Le Moyne, Pierre, 85
L'Enfant, Pierre Charles, 166
Lewis and Clark Expedition, 49, 88, 91, 94, 115, 116, 154
Lexington, Massachusetts, 76
Liliuokalani, Queen, 46
Lincoln, Abraham, 12, 166
Lincoln, Nebraska, 93, 95
Little Rock, Arkansas, 24, 25, 26
Long, Stephen, 94
Long Walk (Navajo Indians), 23
Los Angeles, California, 27, 29
Louisiana, 66–68
Louisiana Purchase, 10
Louisville, Kentucky, 63, 64, 65

Madison, Wisconsin, 159, 161
Maine, 10, 69–71, 167
Manchester, New Hampshire, 99, 100
Manifest Destiny, 11
Marquette, Jacques, 52, 58
Maryland, 9, 72–74
Mason-Dixon Line, 10
Massachusetts, 9, 75–77
Massachusetts Bay Colony, 34, 70
Mayflower, 76
Memphis, Tennessee, 138, 139, 140
Mesabi Range (Minnesota), 55, 82
Mexican War, 11
Michigan, 78–80

Midway, 170
Milwaukee, Wisconsin, 159, 161
mining industry
    coal, 64–65, 115, 127, 128, 157, 163
    copper, 22, 23, 91
    gold, 22, 97, 136
    silver, 22, 49, 97
    turquoise, 23
Minneapolis, Minnesota, 81, 82, 83
Minnesota, 81–83
Mississippi, 84–86
Mississippi River, 8, 84
Missouri, 10, 11, 87–89
Missouri Compromise, 10
Missouri River, 8
Montana, 90–92
Montgomery, Alabama, 15, 16, 17
Montpelier, Vermont, 147
Mormons, 145
Morse, Samuel F. B., 104

Nashville, Tennessee, 138, 139, 140
National Road, 118
Native Americans, 9, 10
Indian reservations, 12, 23, 43, 112, 121
Indian Territory, 10, 121
Nebraska, 12, 93–95
Nevada, 96–98
New Hampshire, 9, 99–101
New Jersey, 9, 102–4
New Mexico, 11, 105–7
New Orleans, Louisiana, 66, 67, 68
New York, 9, 108–10

New York City, New York, 108, 109, 110
Nicolet, Jean, 160
North America's geographic center, 60–61
North Carolina, 9, 111–13
North Dakota, 114–16
Northern Mariana Islands, Commonwealth of the, 168

Oglethorpe, James, 43
Ohio, 117–19
Ohio River, 8, 118
oil industry, 19, 20, 68, 121, 122
Oklahoma, 10, 120–22
Oklahoma City, Oklahoma, 120, 121, 122
Olympia, Washington, 153
Omaha, Nebraska, 93, 94, 95
Oregon, 11, 123–25
organized territories defined, 167

Pacific Ocean, 7, 8
Panic of 1837, 10–11
Pearl Harbor, Hawaii, 13, 47
Penn, William, 127
Pennsylvania, 9, 126–28
Philadelphia, Pennsylvania, 126, 127, 128, 166
Phoenix, Arizona, 21, 22–23
Piedmont Plateau, 8
Pierre, South Dakota, 135, 137
Pike, Zebulon, 31
Pilgrims, 76
Ponce de Léon, Juan, 40
Portland, Maine, 69, 70, 71
Portland, Oregon, 123, 124

Prohibition, 12, 52
Providence, Rhode Island, 129, 130, 131
Puerto Rico, Commonwealth of, 167–68

railroads, 12, 73, 94, 145, 146, 163
Raleigh, North Carolina, 111
Raleigh, Sir Walter, 112
Reconstruction, 12
Revolutionary War, 9, 76, 109, 127
Rhode Island, 9, 10, 129–31
rice production, 25–26
Richmond, Virginia, 150, 151, 152
Rio Grande River, 8
Rocky Mountains, 8
Rodney, Caesar, 37

Sacramento, California, 27, 28
Salem, Oregon, 123, 125
salmon industry, 19
Salt Lake City, Utah, 144, 145, 146
Sante Fe, New Mexico, 105, 106, 107
Sault Ste. Marie, Michigan, 79, 80
Seattle, Washington, 153, 155
Second Seminole War, 40
Sharpsburg battle (Civil War), 73
Sierra Nevada Mountains, 8
silver mining, 22, 49, 97
Slater, Samuel, 130–31
slavery, 10, 11–12, 55, 58, 64
Smith, John, 76
soddies, 61
South Carolina, 9, 132–34
South Dakota, 135–37

Springfield, Illinois, 51, 52, 53
steelmaking industry, 31, 55
St. Lawrence River, 8
St. Louis, Missouri, 87, 88, 89
St. Paul, Minnesota, 81, 82, 83

Tallahassee, Florida, 39, 41
Tennessee, 138–40
Tennessee-Tombigbee Waterway, 17, 85
Tennessee Valley Authority (TVA), 139
Territories of the United States, 167–70
Territory of Guam, 168–69
Terry, Eli, 34
Texas, 11, 141–43, 167
textile industry, 34, 76, 100, 130–31, 134
13th Amendment, 37
Topeka, Kansas, 60, 62
Trail of Tears (Cherokee Indians), 43, 112, 121
transcontinental railroad, 12, 94, 145, 146, 163
Trenton, New Jersey, 102, 104
Tucson, Arizona, 21, 22–23
turquoise mining, 23

Underground Railroad, 11, 58, 64
unincorporated territories defined, 167
United States of America, 7–14
unorganized territories defined, 167
Utah, 11, 12, 144–46

Vancouver, George, 46
Vermont, 147–49, 167
Verrazano, Giovanni da, 103
Vicksburg, Mississippi, 84, 85, 86

Vietnam War, 13
Vikings, 70, 76
Virginia, 9, 150–52
Virgin Islands of The United States, 169
Voting Rights Act, 16–17

Wake Island, 170
Washington, 11, 153–55
Washington, D.C., 11, 73, 165–66
Washington, George, 103, 165
West Virginia, 156–58, 167
wheat production, 61, 82, 115
Whitney, Eli, 34
Wichita, Kansas, 60, 61, 62
Wilmington, Delaware, 36, 37, 38
Wisconsin, 159–61
World War I, 12
World War II, 13
Wyoming, 162–64

Yellowstone National Park, 163, 164
Young, Brigham, 145, 146